United States Nuclear Regulatory Commission

Protecting People and the Environment

NUREG-1940

RASCAL 4: Description of Models and Methods

Office of Nuclear Security and Incident Response

AVAILABILITY OF REFERENCE MATERIALS
IN NRC PUBLICATIONS

United States Nuclear Regulatory Commission

Protecting People and the Environment

NUREG-1940

RASCAL 4: Description of Models and Methods

Manuscript Completed: September 2012
Date Published: December 2012

Prepared by
J.V. Ramsdell, Jr.[a]
G.F. Athey[b]
S.A. McGuire[c]
L.K. Brandon[d]

[a]Pacific Northwest National Laboratory
P.O. Box 999
Richland, WA 99352

[b]Athey Consulting
P.O. Box 178
Charles Town, WV 25414-0178

[c]U.S. Nuclear Regulatory Commission (Retired)

[d]Office of Nuclear Security and Incident Response
U.S. Nuclear Regulatory Commission
Washington, DC 20555-0001

Office of Nuclear Security and Incident Response

ABSTRACT

RASCAL 4 is a significant advancement in the U.S. Nuclear Regulatory Commission's emergency response consequence assessment tools. RASCAL 4 includes improvements in the models and methods related to source term calculations, atmospheric dispersion and deposition, and dose calculations. Changes to the user interface will facilitate data entry, processing, and analysis. This report describes the models and methods that are included in RASCAL 4, and it describes the consequence assessment implications of changes in models and methods from RASCAL 3.

CONTENTS

LIST OF FIGURES

ix

LIST OF TABLES

ACKNOWLEDGEMENTS

We would like to thank all the RASCAL users who have provided feedback on the models. The software continues to grow and improve because of your efforts and ideas.

INTRODUCTION

RASCAL 4 consists of several modules. Consequence assessments for nuclear power plants use five of the modules. The code invokes four of these modules when the user selects "Source Term to Dose" on the opening screen. The first module calculates the time-dependent atmospheric release source term. The atmospheric release source term is the rate at which radioactive material is released to the environment. It also includes other information that defines how the release takes place. The second and third modules perform the atmospheric transport, dispersion, and deposition calculations and the dose calculations The fourth module is used to create the meteorological data file used by the atmospheric transport, dispersion, and deposition modules. The fifth module is used for intermediate-phase dose calculations based on field measurements. Uranium fuel cycle consequence assessments use the sixth module.

This document presents the technical bases for these six modules. The first three chapters present the technical bases, including assumptions, for estimating source terms. In order, the chapters cover nuclear power plant source terms, spent fuel storage facility source terms, and fuel cycle and materials source terms. Chapter 4 presents the technical bases and assumptions associated with the atmospheric and dose calculations involving radionuclides. Chapter 5 presents the technical bases and assumptions for the atmospheric transports and dispersion calculations involving uranium hexafluoride (UF_6). Chapter 6 presents technical bases and assumptions related to processing meteorological data for use by the atmospheric models. Finally, Chapter 7 presents technical bases and assumptions related to the calculation of intermediate-phase doses based on field measurements. Each chapter discusses significant changes from previous versions of RASCAL and verification and validation.

All the RASCAL 4 modules, except for the meteorological data processing module and the UF_6 atmospheric module, calculate radioactive decay and daughter ingrowth. A common approach is used throughout RASCAL 4. Appendix A describes this approach in detail. Appendix A has (1) a table of parent radionuclides and their implicit daughters, (2) a table of the composite dose conversion factors for those radionuclides with implicit daughters, (3) a table of short-lived radionuclides that do not appear explicitly in RASCAL 4, and (4) a table of the radionuclides that RASCAL 4 treats explicitly. This last table includes the decay chain for each isotope.

1. NUCLEAR POWER PLANT SOURCE TERM CALCULATIONS

This chapter describes how the "Source Term to Dose" module in RASCAL 4 calculates the time-dependent source term for nuclear power plant accidents. The methods that the RASCAL 4 source term calculations use for nuclear power plant accidents are based largely on the methods described in NUREG-1228, "Source Term Estimation during Incident Response to Severe Nuclear Power Plant Accidents," (McKenna and Giitter, 1988). Various aspects of the source term estimation methodology, including release timing, have been modified to account for the accident source term insights in NUREG-1465, "Accident Source Terms for Light-Water Nuclear Power Plants, Final Report," (Soffer et al., 1995).

1.1 Nuclear Power Plant Parameters

Before describing the detailed methods that RASCAL 4 uses to calculate specific source term types, this report first describes the nuclear power plant parameters that are used in the calculations.

1.1.1 Core Inventories

For nuclear power plant source terms based on core damage, Table 1-1 shows the radionuclide inventories that RASCAL 4 assumes are in the reactor core. The normalized core inventories (curies per megawatts thermal) in the table are based on calculations made by the U.S. Nuclear Regulatory Commission (NRC) staff in December 2003 using the SAS2H control module of SCALE (Standardized Computer Analyses for Licensing Evaluation), Version 4.4a. SAS2H uses the point depletion code ORIGEN-S to compute time-dependent concentrations of a large number of radionuclides. The calculations were done for a single fuel assembly with a burnup of 38,585 megawatt days per metric ton of uranium (MWd/MTU). The core contained 193 assemblies and had a power level of 3,479 megawatts thermal (MWt). The enrichment of the assembly was 4.0 weight percent uranium-235. Normalization was done by multiplying the SAS2H-calculated inventory for a single assembly by 193 assemblies per core and then dividing by 3,479 MWt.

Table 1-1 contains radionuclides, such as cesium-137* (Cs-137*), that include a * symbol. RASCAL 4 assumes that these radionuclides are present in secular equilibrium with short-lived daughters. For example, RASCAL 4 assumes that Cs-137* includes barium-137m (Ba-137m), which has a half-life of 2.552 minutes. Numerical procedures in the atmospheric transport modules preclude the explicit representation of radionuclides with short half-lives. As a result, the inventories do not include radionuclides with half-lives of 10 minutes or less. However, where appropriate, the inventories implicitly include short-lived daughters with their longer lived parents for dose calculations. Appendix A includes a full list of radionuclides that have implicit daughters in RASCAL and another list of short-lived radionuclides that are not included in the RASCAL 4 radionuclide library. The implicit daughters do not alter the basic characteristics of the parent radionuclides other than increasing dose conversion factors.

Table 1-1 Assumed Core Inventory during Operation for Low-Enriched Uranium Fuel

NUCLIDE	CORE INVENTORY (Ci/MWt)	NUCLIDE	CORE INVENTORY (Ci/MWt)	NUCLIDE	CORE INVENTORY (Ci/MWt)
Ba-139	4.74E+04	La-141	4.33E+04	Te-127	2.36E+03
Ba-140	4.76E+04	La-142	4.21E+04	Te-127m	3.97E+02
Ce-141	4.39E+04	Mo-99	5.30E+04	Te-129	8.26E+03
Ce-143	4.00E+04	Nb-95	4.50E+04	Te-129m	1.68E+03
Ce-144*	3.54E+04	Nd-147	1.75E+04	Te-131m	5.41E+03
Cm-242	1.12E+03	Np-239	5.69E+05	Te-132	3.81E+04
Cs-134	4.70E+03	Pr-143	3.96E+04	Xe-131m	3.65E+02
Cs-136	1.49E+03	Pu-241	4.26E+03	Xe-133	5.43E+04
Cs-137*	3.25E+03	Rb-86	5.29E+01	Xe-133m	1.72E+03
I-131	2.67E+04	Rh-105	2.81E+04	Xe-135	1.42E+04
I-132	3.88E+04	Ru-103	4.34E+04	Xe-135m	1.15E+04
I-133	5.42E+04	Ru-105	3.06E+04	Xe-138	4.56E+04
I-134	5.98E+04	Ru-106*	1.55E+04	Y-90	2.45E+03
I-135	5.18E+04	Sb-127	2.39E+03	Y-91	3.17E+04
Kr-83m	3.05E+03	Sb-129	8.68E+03	Y-92	3.26E+04
Kr-85	2.78E+02	Sr-89	2.41E+04	Y-93	2.52E+04
Kr-85m	6.17E+03	Sr-90	2.39E+03	Zr-95	4.44E+04
Kr-87	1.23E+04	Sr-91	3.01E+04	Zr-97*	4.23E+04
Kr-88	1.70E+04	Sr-92	3.24E+04		
La-140	4.91E+04	Tc-99m	4.37E+04		

RASCAL 4 adjusts the inventory of radionuclides that have half-lives that exceed 1 year to account for burnup. Equation 1-1 is used to calculate the inventory for the core-average burnup, I_{ACTUAL}, for nuclides with half-lives of more than 1 year. Inventories of nuclides with half-lives of less than 1 year are not adjusted for burnup because the activities for these nuclides are more closely related to reactor power than they are to burnup.

$$I_{ACTUAL} = I_{38,585} \times \frac{BURNUP_{ACTUAL}}{38,585 \text{ MWd/MTU}} \tag{1-1}$$

If the reactor is shut down before the start of the release, the radionuclide inventories are adjusted to account for radiological decay and ingrowth. In addition, at the end of each time step, RASCAL adjusts the activities of the nuclides present to account for radiological decay and ingrowth. When the activity released during a source term time step decreases to less than 1 becquerel ((Bq) (2.7×10^{-11} curies (Ci)) for a nuclide, the activity is set to zero.

1.1.2 Coolant Inventories

RASCAL 4 uses coolant inventories for some accident types. Table 1-2 lists the nuclide concentrations that RASCAL 4 uses for normal coolant. Those normal coolant concentrations are taken from American National Standard Institute/American Nuclear Society (ANSI/ANS) 18 1-1999, "Radioactive Source Term for Normal Operation of Light-Water Reactors,". Coolant concentration for boiling-water reactors (BWRs) were taken from Table 5 in ANSI/ANS 18.1-1999. The values for reactor steam were used for the noble gases, and the values for reactor water were used for all the other radionuclides. In cases in which a parent/daughter pair is shown, only the parent was included. Coolant concentrations for pressurized-water reactors (PWRs) were taken from Table 6 in ANSI/ANS 18.1-1999. This table was selected over Table 7 in ANSI/ANS 18.1-1999 because it provides values for reactors with U-tube steam generators (SGs) (the most common type). The values for reactor coolant were used from the table.

During steady-state conditions, iodine and other fission products may escape from fuel rods that have cladding defects and enter the reactor coolant system (RCS). The rate of escape of fission products is low because the internal pressure in the fuel rod is balanced with the coolant pressure outside the fuel rod during steady-state conditions. In addition, the RCS purification cleanup continually removes fission products that do escape into the RCS. As a result, the equilibrium concentration of fission products in the coolant remains low.

However, if a reactor transient causes the pressure of the RCS to decrease rapidly, the escape rate from fuel rods can increase and cause a temporary increase, or "spike," in the coolant concentrations. Coolant water may also enter fuel rods through cladding defects. If the RCS pressure suddenly decreases, this water could leach off iodine and cesium salts that are deposited on the inner cladding surfaces, thus increasing the iodine and cesium available for escape during the transient.

RASCAL 4 can also calculate an inventory for "spiked" coolant. It assumes a spiking factor to increase the concentrations of halogens (iodine) and alkali metals (cesium) in the coolant. RASCAL 4 uses a default spiking factor of 30, but the user can enter a different value.

Following reactor shutdown, RASCAL 4 uses the decay and ingrowth methodology described in Appendix A to adjust the coolant inventories.

Table 1-2 Radionuclide Concentrations in Reactor Coolant*

NUCLIDE	PWR COOLANT CONCENTRATION (Ci/g)	BWR COOLANT CONCENTRATION (Ci/g)	NUCLIDE	PWR COOLANT CONCENTRATION (Ci/g)	BWR COOLANT CONCENTRATION (Ci/g)
Ag-110m*	1.3E-09	1.0E-12	Nb-95	2.8E-10	0.0E+00
Ba-140	1.3E-08	4.0E-10	Ni-63	0.0E+00	1.0E-12
Br-84	1.6E-08	0.0E+00	Np-239	2.2E-09	8.0E-09
Ce-141	1.5E-10	3.0E-11	P-32	0.0E+00	4.0E-11
Ce-143	2.8E-09	0.0E+00	Rb-88	1.9E-07	0.0E+00
Ce-144*	4.0E-09	3.0E-12	Rb-89	0.0E+00	5.0E-09
Co-58	4.6E-09	1.0E-10	Ru-103	7.5E-09	2.0E-11
Co-60	5.3E-10	2.0E-10	Ru-106*	9.0E-08	3.0E-12
Cr-51	3.1E-09	3.0E-09	Sr-89	1.4E-10	1.0E-10
Cs-134	3.7E-11	3.0E-11	Sr-90	1.2E-11	7.0E-12
Cs-136	8.7E-10	2.0E-11	Sr-91	9.6E-10	4.0E-09
Cs-137*	5.3E-11	8.0E-11	Sr-92	0.0E+00	1.0E-08
Cs-138	0.0E+00	1.0E-08	Tc-99m	4.7E-09	0.0E+00
Cu-64	0.0E+00	3.0E-09	Te-129	2.4E-08	0.0E+00
Fe-55	1.2E-09	1.0E-09	Te-129m	1.9E-10	4.0E-11
Fe-59	3.0E-10	3.0E-11	Te-131	7.7E-09	0.0E+00
H-3	1.0E-06	1.0E-08	Te-131m	1.5E-09	1.0E-10
I-131	2.0E-09	2.2E-09	Te-132	1.7E-09	1.0E-11
I-132	6.0E-08	2.2E-08	W-187	2.5E-09	3.0E-10
I-133	2.6E-08	1.5E-08	Xe-131m	7.3E-07	3.3E-12
I-134	1.0E-07	4.3E-08	Xe-133	2.9E-08	1.4E-09
I-135	5.5E-08	2.2E-08	Xe-133m	7.0E-08	4.9E-11
Kr-83m	0.0E+00	5.9E-10	Xe-135	6.7E-08	3.8E-09
Kr-85	4.3E-07	4.0E-12	Xe-135m	1.3E-07	4.4E-09
Kr-85m	1.6E-08	1.0E-09	Xe-138	6.1E-08	1.5E-08
Kr-87	1.7E-08	3.3E-09	Y-91	5.2E-12	4.0E-11
Kr-88	1.8E-08	3.3E-09	Y-91m	4.6E-10	0.0E+00
La-140	2.5E-08	0.0E+00	Y-92	0.0E+00	6.0E-09
Mn-54	1.6E-09	3.5E-11	Y-93	4.2E-09	4.0E-09
Mn-56	0.0E+00	2.5E-08	Zn-65	5.1E-10	1.0E-10
Mo-99	6.4E-09	2.0E-09	Zr-95	3.9E-10	8.0E-12
Na-24	4.7E-08	2.0E-09			

*Reference: ANSI/ANS 18.1-1999.

1.1.3 Reactor Coolant System Water Mass

The RASCAL 4 facility database stores the mass of water (kilograms) in the RCS for each plant. The RCS coolant system water masses were estimated for each reactor. Table 1-3 provides the mass of water in the reactor vessel or in the RCS in kilograms for the type of reactor listed (this information was taken from ANSI/ANS 18.1-1999).

Table 1-3   Reference Reactor Water Mass*

3,400 MWt REACTOR	MASS OF WATER IN THE REACTOR VESSEL (BWR) OR IN THE RCS (PWR)	SOURCE TABLE
BWR	1.7E+5 kg	1
PWR with U-tube SG	2.5E+5 kg	2
PWR with once-through SG	2.5E+5 kg	3

*Source: ANSI/ANS 18.1-1999.

For each reactor, the actual licensed power (MWt) was divided by the reference power (3,400 MWt) and then multiplied by the mass of the water (kilograms from the table above) to estimate the coolant mass.

For example, Beaver Valley Power Station, Unit 1 (a PWR), had a licensed power of 2,652 MWt. Thus, in RASCAL, the mass of water in the Beaver Valley Unit 1 RCS primary coolant is

$$2,652 / 3,400 \times 2.5 \times 10^5 = 1.95 \times 10^5 \text{ kilograms (kg)}.$$

This method produces only the approximate RCS primary coolant mass. Plant technical specifications include more accurate site-specific values. However, the doses associated with releases from the RCS are small compared with the doses from releases of core activity.

1.1.4 Reactor Containment Air Volumes

The containment air volumes in the RASCAL 4 database were taken from plant final safety analysis reports (FSARs). For PWRs, the volumes are the total containment volumes. For BWRs, the volumes are the drywell volume plus the wetwell air space volume. Units in the database are in cubic feet.

1.1.5 Reactor Power Levels

The RASCAL 4 database lists reactor power levels in units of megawatt thermal. These units represent the maximum power at which the reactor is allowed to operate. This value is used as the default value for average reactor power, but the user may change the value from the RASCAL 4 user interface. These values were originally taken from the NRC Information Digest. The RASCAL developers have updated them for RASCAL 4 so that they are current with NRC-approved power upgrades as of March 2012. (See the NRC Web site at www.nrc.gov/reactors/operating/licensing/power-uprates.html.)

1.1.6 Fuel Burnup

The RASCAL 4 database contains two fuel burnup numbers. The first is the core average burnup (megawatt days per metric ton of uranium) for each reactor. A default value of 30,000 MWd/MTU is used in the database for all reactors. This value represents the average burnup of a core that is roughly two-thirds of the way to the end of core life, assuming typical current fuel management practices. The value changes with time and with the mix of old and new fuel in the core. The user can change the value if more information is available, but usually such a change will not significantly change the calculated projected doses. This burnup number is used to adjust the available inventory of radionuclides with half-lives greater than 1 year (Section 1.1.1).

The second burnup number is the peak rod burnup. The peak-rod burnup is used as the burnup of fuel to be sent to the spent-fuel pool for long-term storage.. A value of 50,000 MWd/MTU is used in the database. Again, the user may change the value if a better number is available. The spent fuel burnup is used to generate source terms for spent fuel accidents using the method in Section 1.1.1.

1.1.7 Number of Assemblies in the Core

The RASCAL 4 database contains the number of fuel assemblies in each reactor core. The values are taken from plant FSARs. Only spent fuel accident calculations make use of these numbers when estimating the source term activity for a fuel assembly (Chapter 2).

1.1.8 Design Pressure

The RASCAL 4 database includes a design pressure for each reactor containment. The values are in pounds per square inch. The design pressures are taken from plant FSARs. The user cannot change this value from the RASCAL 4 user interface, but the case summary does display it. RASCAL 4 calculations do not use the design pressure, but the user can compare the actual containment pressure to the design pressure to determine whether the actual leak rate is near or below the design leak rate.

1.1.9 Design Leak Rate

The RASCAL 4 database includes a design leak rate for each reactor containment. The values are in percent of containment volume per day at design pressure. The design leak rates are taken from plant FSARs. The user cannot change this value from the RASCAL 4 user interface. The design leak rate is the default containment leak rate, but the RASCAL 4 user may select a leak rate more appropriate for a particular accident.

1.1.10 Boiling-Water Reactor Stack Heights

Some BWRs have a tall stack through which releases from the standby gas treatment system (SBGTS) can be routed. The heights of these BWR stacks are included in the RASCAL 4 facility database and are taken from plant FSARs. The user cannot change the values from the RASCAL 4 user interface. The values are used to set the release height for releases through the SBGTS.

1.1.11 Pressurized-Water Reactor Steam Generator Water Mass

The RASCAL 4 database includes the mass of water in a PWR steam generator. A value of 42,184 kg (93,000 pounds) is used for all reactors. Reactor-specific values are not included. This value represents a typical value derived from accident analyses. The water volume fluctuates rapidly and is not constant during normal operation.

1.2 Source Term Types

1.2.1 Basic Method Used To Calculate Source Terms

A source term is defined as the activities of each radionuclide released to the environment as a function of time. The basic method used to calculate a source term is to divide the nuclear power plant into compartments and to then calculate the activities entering the compartment and the activities being removed from the compartment during time steps of fairly short duration. The time steps generally have a 15-minute duration.

As an example, consider a loss-of-coolant accident at a PWR after reactor shutdown in which fuel is damaged and radionuclides are released to the containment and then to the atmosphere. The first compartment is the fuel. RASCAL 4 will first calculate the release from the fuel to the containment atmosphere. Because the reactor is shutdown, it is not producing any new fission products. Radiological decay and the release of these decay products to the containment deplete the radionuclide inventory of the fuel during each time step. In addition, ingrowth of some radionuclides will occur in the fuel from the radiological decay of their parents. Appendix A describes the RASCAL 4 methods and assumptions for calculating decay and ingrowth.

The second compartment is the containment. The activity entering the containment during a time step is the activity released from the fuel during that time step. Radiological decay, removal processes (e.g., removal by containment sprays), and leakage to the environment remove the activity from the containment atmosphere during the time step.

Time steps may vary in length. A source term time step starts whenever the user changes any of the time-dependent data or every 15 minutes, whichever occurs first. Time steps may be no less than 1 minute and must be an integral number of minutes. Before passing the source term to the atmospheric transport model, RASCAL 4 converts the source term time steps into 15-minute time steps because the atmospheric transport models require that regularity.

The remainder of this chapter describes in detail how RASCAL 4 calculates the time-dependent source term for various accident types.

1.2.2 Time Core Is Uncovered Source Term

Perhaps the most powerful and important source term type that RASCAL 4 calculates is based on the time that the core is uncovered. Almost all of the radioactivity at a nuclear power plant is contained in fuel rods. A large release is not possible unless many fuel rods are substantially damaged. Loss of water from the primary coolant system that leaves the reactor core uncovered is the only way that this large release can reasonable occur. If a user estimates how long a reactor core will be uncovered by water, RASCAL 4 can estimate the amount of core damage that will occur and, from that estimate, the activity of each fission product nuclide that will be released from the core.

When a RASCAL 4 user specifies how long the core is uncovered, RASCAL 4 will estimate how much core damage will occur based on the damage timings in Table 1-4 for BWRs and Table 1-5 for PWRs. (Tables 1-4 and 1-5 are taken from Tables 3-12 and 3-13 in NUREG-1465.) For example, if a BWR or PWR core is uncovered for 15 or 30 minutes, the estimated damage is 50- or 100-percent cladding failure, respectively. If a BWR core is uncovered for 1 hour, the estimated damage will be 100-percent cladding failure plus 33-percent core melt.

Table 1-4 BWR Event Timings and Fraction of Core Activity Inventory Released*

NUCLIDE GROUP	BWR CORE INVENTORY RELEASE FRACTION		
	Cladding Failure (Gap Release Phase) (0.5-hour duration)	Core Melt Phase (In-Vessel Phase) (1.5-hour duration)	Postvessel Melt-Through Phase (Ex-Vessel Phase) (3.0-hour duration)
Noble gases (Kr, Xe)	0.05	0.95	0
Halogens (I, Br)	0.05	0.25	0.30
Alkali metals (Cs, Rb)	0.05	0.20	0.35
Tellurium group (Te, Sb, Se)	0	0.05	0.25
Barium, strontium (Ba, Sr)	0	0.02	0.1
Noble metals (Ru, Rh, Pd, Mo, Tc, Co)	0	0.0025	0.0025
Cerium group (Ce, Pu, Np)	0	0.0005	0.005
Lanthanides (La, Zr, Nd, Eu, Nb, Pm, Pr, Sm, Y, Cm, Am)	0	0.0002	0.005

*Reference: Table 3-12 from NUREG-1465.

The fractions shown in these tables are for the particular phase. They are not cumulative. Thus, the total fraction of core inventory released in a vessel melt-through accident is the sum of the fractions for cladding failure, core melt, and vessel melt-through.

The data in Tables 1-4 and 1-5 are the result of an expert elucidation process that considered a range of severe accident sequences. These release fractions do not cover all potential severe accident sequences, nor do they represent any particular accident sequence. However, the timings in Tables 1-4 and 1-5 for the start of each fuel damage state were based on the accident sequence that could lead to the earliest fuel failures. The timings and release fractions in Tables 1-4 and 1-5 are essentially based on a large-break loss-of-coolant accident with the reactor at full power and without the operation of emergency core cooling systems. This situation leads to a very rapid uncovering of the core.

However, if a small break in the RCS occurs or if the emergency core cooling systems initially operated successfully, the core will remain covered while the rate of decay heat production decreases. At lower decay heat production rates, the duration of each release phase is likely to increase, and Tables 1-4 and 1-5 may well have overestimated the release fractions during each release phase. However, RASCAL 4 does not adjust its releases to account for that situation. Users of the "time core is uncovered" source term option should understand that RASCAL 4 is likely to overestimate the speed and magnitude of the release and thus also overestimate the projected radiological doses. RASCAL 4 users should inform decisionmakers of that fact.

Table 1-5 PWR Event Timings and Fraction of Core Activity Inventory Released*

NUCLIDE GROUP	PWR CORE INVENTORY RELEASE FRACTION		
	Cladding Failure (Gap Release Phase) (0.5-hour duration)	Core Melt Phase (In-Vessel Phase) (1.3-hour duration)	Postvessel Melt-Through Phase (Ex-Vessel Phase) (2.0-hour duration)
Noble gases (Kr, Xe)	0.05	0.95	0
Halogens (I, Br)	0.05	0.35	0.25
Alkali metals (Cs, Rb)	0.05	0.25	0.35
Tellurium group (Te, Sb, Se)	0	0.05	0.25
Barium, strontium (Ba, Sr)	0	0.02	0.1
Noble metals (Ru, Rh, Pd, Mo, Tc, Co)	0	0.0025	0.0025
Cerium group (Ce, Pu, Np)	0	0.0005	0.005
Lanthanides (La, Zr, Nd, Eu, Nb, Pm, Pr, Sm, Y, Cm, Am)	0	0.0002	0.005

*Reference: Table 3-13 from NUREG-1465.

For PWRs, the time that the core is uncovered should be the time that the coolant drops below the top of the active fuel. At this level, cladding failure will begin. The gap activity in each fuel rod is released suddenly when the cladding fails at some location caused by overpressure. The rods near the center of the core will fail earliest; additional rods will fail as the core continues to heat. This process takes about 30 minutes.

For BWRs, the cladding damage does not start until the water uncovers about one-third of the way down the fuel element. Before that time, the boiling water below will keep the fuel cool enough to prevent the cladding from melting.

For calculations using the "time core is uncovered" source term type, RASCAL 4 will first calculate the activity released from the fuel either to the containment atmosphere or to the coolant as appropriate for the release pathway that the user has selected. Equation 1-2 is as follows:

$$A_i(k) = I_i \times AF_i(k), \tag{1-2}$$

where:

I_i = the core inventory of radionuclide i
$AF_i(k)$ = the available fraction of the inventory of radionuclide i available for release from the fuel during time step k

To illustrate how RASCAL 4 calculates the $AF_i(k)$, consider a PWR for the fourth 15-minute time step (45 minutes to 60 minutes), during which time the reactor is entirely in the core melt phase. During the core melt phase, 95 percent of the noble gases would be released over 1.3 hours according to Table 1-4. Equation 1-3 is used to calculate the *AF* for the release of noble gas activity from the fuel to the containment, $AF_{ng}(4)$, during that fourth 0.25-hour-duration time step, as follows:

$$AF_{ng}(4) = 0.95 \frac{0.25 \ hour}{1.3 \ hour},$$ (1-3)

If the user enters a time at which the core is recovered with water, the calculation assumes that core damage stops along with the release of material from the core at that time.

1.2.3 Specified Core Damage Endpoint Source Term

The user can specify the expected maximum damage by selecting a core-damage state directly. The state that the user selects will establish the source term. The user can select (1) no core damage (normal coolant activity), (2) increased fuel pin leakage with spiked coolant activity, or (3) 1- to 100-percent cladding failure.

The user also selects the time at which the maximum damage is expected to occur. For example, if the user believes that a maximum damage of 10-percent cladding failure may occur, the user enters the time at which he or she believes that 10 percent of the cladding will have failed.

1.2.3.1 Normal Coolant

For normal coolant releases, RASCAL 4 uses the coolant concentrations from Table 1-2 that are decayed from the time of shutdown to the time entered as the point of maximum damage. The inventory I_i is the concentration of radionuclide *i* times the total coolant mass. The available fraction for release is the mass of coolant escaping during the time step divided by the total coolant mass.

1.2.3.2 Spiked Coolant

Spiked coolant may occur following reactor shutdown, startup, rapid power change, and RCS depressurization. Rapid increases in the iodine and other fission-product concentrations in the coolant as high as 3 orders of magnitude may occur. The default spiking factor is 30, but the user can select a spiking factor from 1 to 1,000.

For spiked coolant releases, RASCAL 4 uses the coolant concentrations (Table 1-2). The concentration of all halogens (iodine) and alkali metals (cesium) in the coolant are multiplied by the spiking factor.

For both normal and spiked coolant releases, only the steam generator and containment bypass release pathways are available. The user must specify the mass leak rate at which coolant escapes the RCS. Generally, the user can assume that the leak rate is the same as the makeup flow needed to maintain the water level.

1.2.3.3 Cladding Failure

For cladding failure, RASCAL 4 uses the AFs determined from Tables 1-3 and 1-4. For example, if the user entered 4-percent cladding failure for a BWR, the iodine release from the fuel would be as follows: core inventory of iodine times 0.04 (fraction of cladding that failed) times 0.05 (the halogen AF for 100-percent cladding failure in Table 1-3).

The user cannot select core melt or vessel melt-through as a specified core damage endpoint as he or she could in previous versions (Version 3.0.4 or earlier) of RASCAL. For accidents proceeding to core melt, the user interface screen tells the user to use the time core is uncovered source term type. For accidents with core melt, the timing of the release would not be at all realistic using the specified core damage endpoint source term option. In addition, although the user may have a relatively good idea of when core damage may begin, he or she may have less knowledge of when the maximum damage will occur. For these reasons, the use of the time core is uncovered source term type will give more realistic results for accidents that are expected to proceed into core melt.

RASCAL 4 calculates how much activity will be released with the escaping coolant during the first 15-minute time step. For the second time step, RASCAL 4 decreases the concentration of radionuclides in the coolant to account for what has escaped. RASCAL 4 assumes that the makeup water that is being added to the primary system is clean water. Using the common RASCAL approach (Appendix A) of adjusting the concentrations during each time step accounts for radioactive decay and ingrowth.

The two release pathways that can release coolant are a steam generator tube rupture and a containment bypass pathway. The user must specify the leak rate at which coolant escapes the RCS. Generally, the user can assume that the leak rate is the same as the makeup flow needed to maintain the water level. This chapter later discusses the reduction factors that are applied before the release to the environment.

1.2.4 Containment Radiation Monitor Source Term

RASCAL 4 can use containment radiation monitor readings to estimate source terms that occur through the containment leakage release pathway. The user enters containment radiation monitor readings and the times of the readings. The entry of multiple readings allows the modeling of core damage that is progressing with time.

Figures 1-1–1-5 show the expected containment radiation monitor readings related to radionuclides in the containment atmosphere from coolant or core damage. The figures show the calculated monitor readings at 1 hour and 24 hours after shutdown. These figures are taken from Figures A.5 through A.12 in NUREG/BR-0150, "Response Technical Manual (RTM)-96 (McKenna et al., 1996). The bars in these figures represent the calculated containment radiation monitor readings for 1 to 100 percent of the labeled core-damage state.

The data in these figures were calculated for a reactor power of 3,000 MWt. RASCAL 4 scales the monitor reading entered by the user to account for the difference in the 3,000 MWt reactor power used to produce the figure and the actual reactor power. Equation 1-4 is used to calculate this scaled monitor reading R, as follows:

$$R = \frac{3000 \times MR}{Power}, \qquad (1\text{-}4)$$

where:

MR = the containment monitor reading entered
$Power$ = the reactor power, MWt.

To estimate a source term from the scaled monitor reading, RASCAL 4 first determines which figure should be used based on the containment type and, for BWRs, the monitor location (drywell or wetwell). Next, RASCAL 4 determines whether the data for containment "sprays on" or "sprays off" should be used.

RASCAL 4 then adjusts the data in the figure for the time between shutdown and the monitor reading. If the holdup time entered is less than 1 hour, RASCAL 4 uses the data for 1 hour without adjustment. If the holdup time is greater than 24 hours, RASCAL 4 uses the data for 24 hours without adjustment. If the holdup time is between 1 hour and 24 hours, RASCAL 4 does a linear interpolation to calculate a new 1- to 100-percent bar for actual holdup time (the time between the shutdown time and the time of the monitor reading).

Figures 1-1–1-5 show the containment monitor readings for (1) normal coolant, (2) spiked coolant, (3) cladding failure, and (4) core melt. If a containment radiation monitor reading exceeds the value of 1-percent core melt, RASCAL 4 assumes that core melt has begun and uses a core melt source term.

If the containment radiation monitor reading is less than the value for 1-percent core melt, RASCAL 4 assumes that the core damage state is cladding failure. RASCAL 4 does not use the data in the figures for normal and spiked coolant.

When RASCAL 4 determines that the estimated core-damage state is cladding failure or core melt, it computes the fraction of either state that the reading represents. This percentage cannot be more than 100 percent. Equation 1-5 is used to calculate the percentage P of the selected damage state, as follows:

$$P = 100 \times \frac{R}{P_{1D}},\tag{1-5}$$

where P_{1D} is the meter reading assumed for 1 percent of the core-damage state for a 3,000-MWt reactor.

RASCAL 4 prevents this method from predicting a decrease in core damage over time. Going from a prediction of 50-percent core melt to 10-percent cladding failure would be unrealistic. The user interface will allow actual monitor readings to decrease with time because the interaction with spray status and time after shutdown might result in an increase in core damage. However, the source term model ignores any decrease in damage, and the user is alerted upon detection of the condition.

The user should be very cautious in interpreting RASCAL 4 results based on containment radiation monitor readings because the calculations are subject to large uncertainties and have certain limitations, as follows:

- The model assumes that the containment radiation monitor readings represent the full amount of damage that has occurred. However, if the fission products are delayed in entering the containment, the containment monitor readings may significantly lag behind the amount of damage that has occurred.

- The model also assumes a uniform mixing of fission products in the containment atmosphere. Uneven mixing in containment, such as if steam rises to the top of the dome or if insufficient time prevents uniform mixing, may cause inconsistent readings. If uniform mixing has not yet occurred, the monitor readings may significantly misrepresent the amount of damage that has occurred.

- The model also assumes that an unshielded monitor sees a large fraction of the containment volume. If that is not true, significant error could result. Because the mix is most likely different from that assumed in the calibration of the monitor, the actual reading at the upper end of the scale could differ significantly if a shielded detector is used for the higher radiation measurements.

- Figures 1-1–1-5 represent typical reactor plants. Plant-specific conditions may make differences.

- Figures 1-1–1-5 are appropriate for large-break loss-of-coolant accidents. If a small break occurs, the containment activity may rise very slowly at first, thus causing RASCAL 4 to underestimate the amount of core damage that has occurred.

- Thermal stratification in the containment may affect the results. The containment atmosphere near the containment radiation monitors may not be representative of the containment atmosphere as a whole.

- The containment radiation monitor source term is a lagging indicator of core damage and cannot predict core damage that will occur in the future. Thus, it will be much later in its estimates of projected doses compared to the "time core is uncovered" source term.

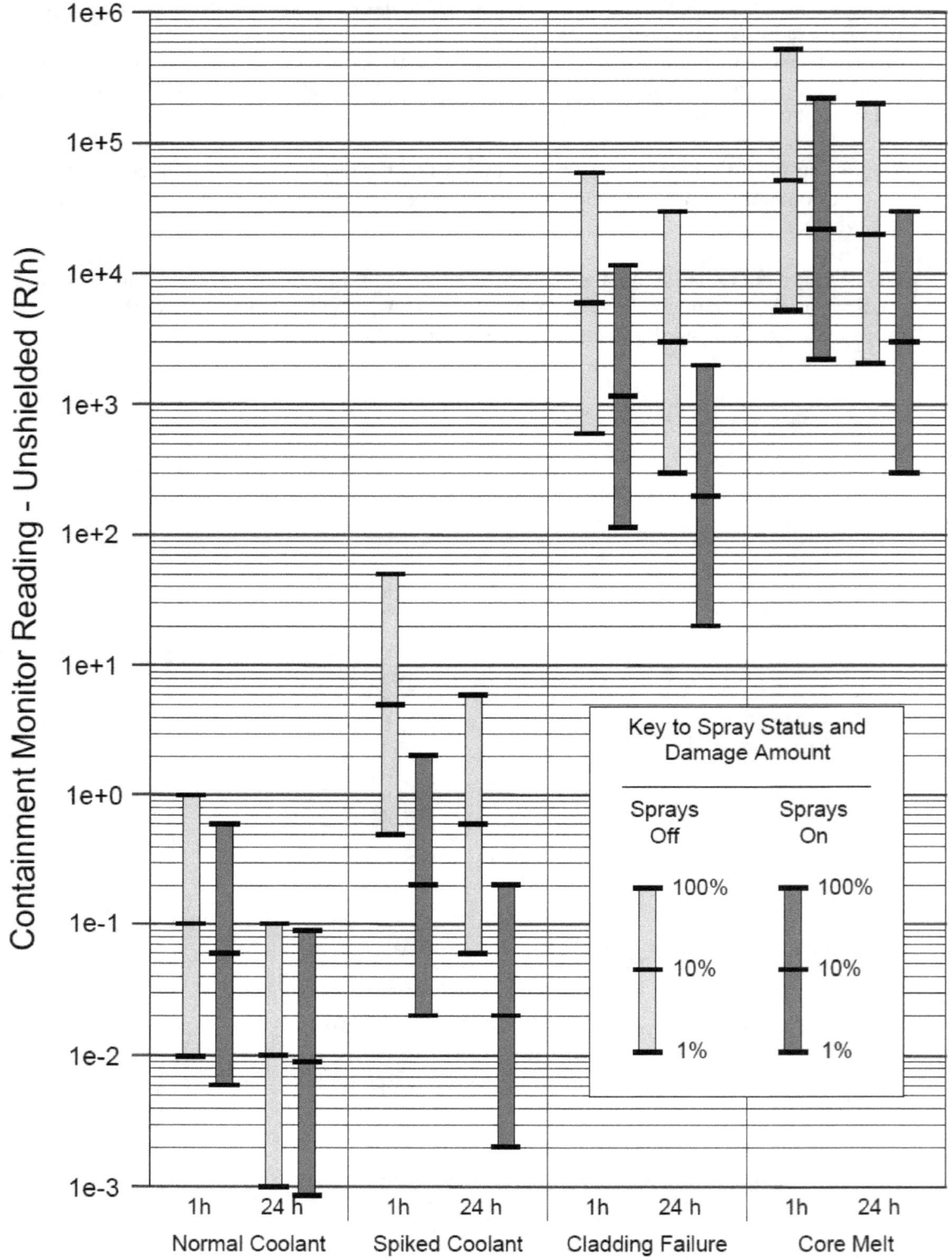

Figure 1-1 PWR containment monitor response

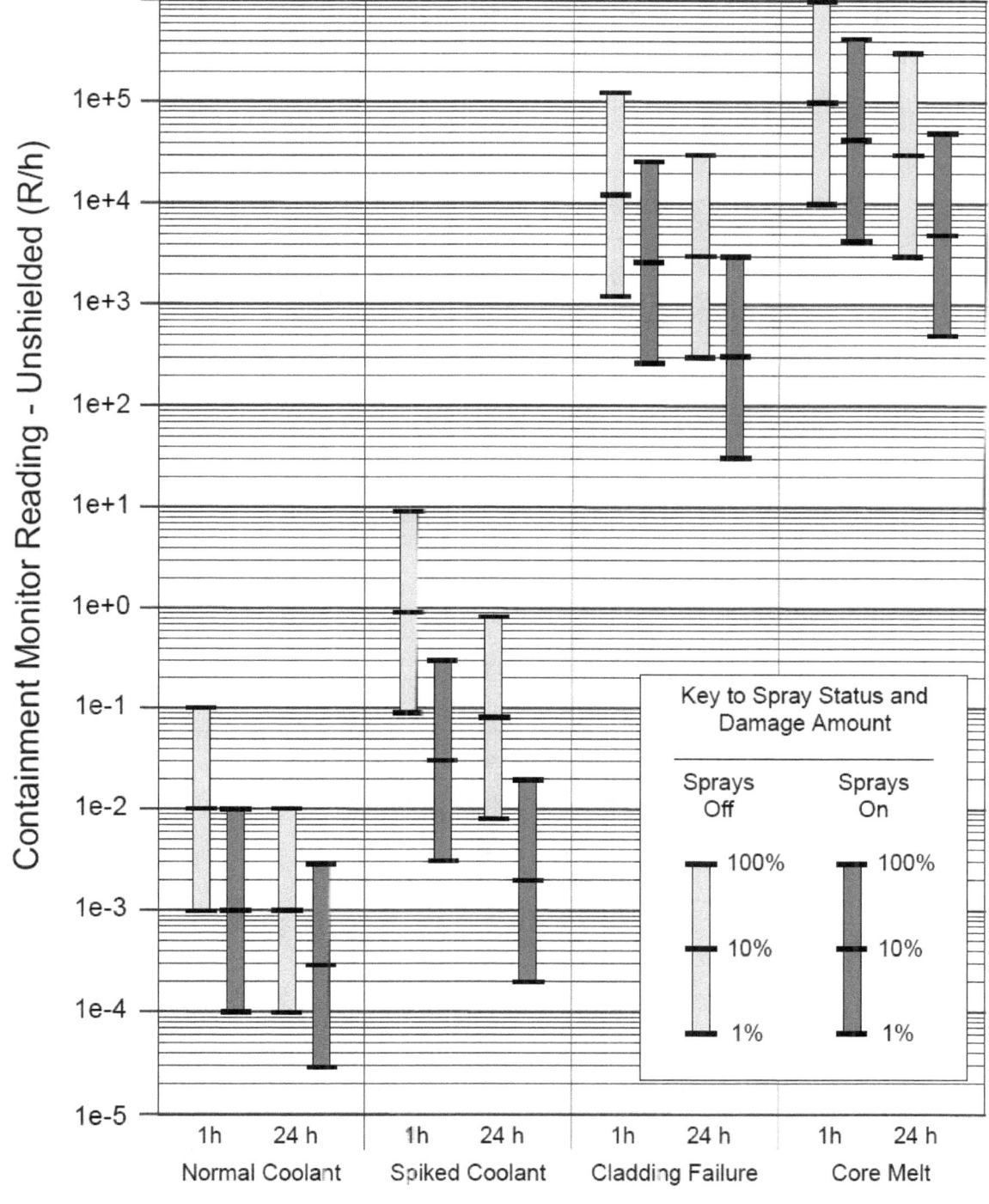

Figure 1-2 BWR Mark I and Mark II drywell containment monitor response

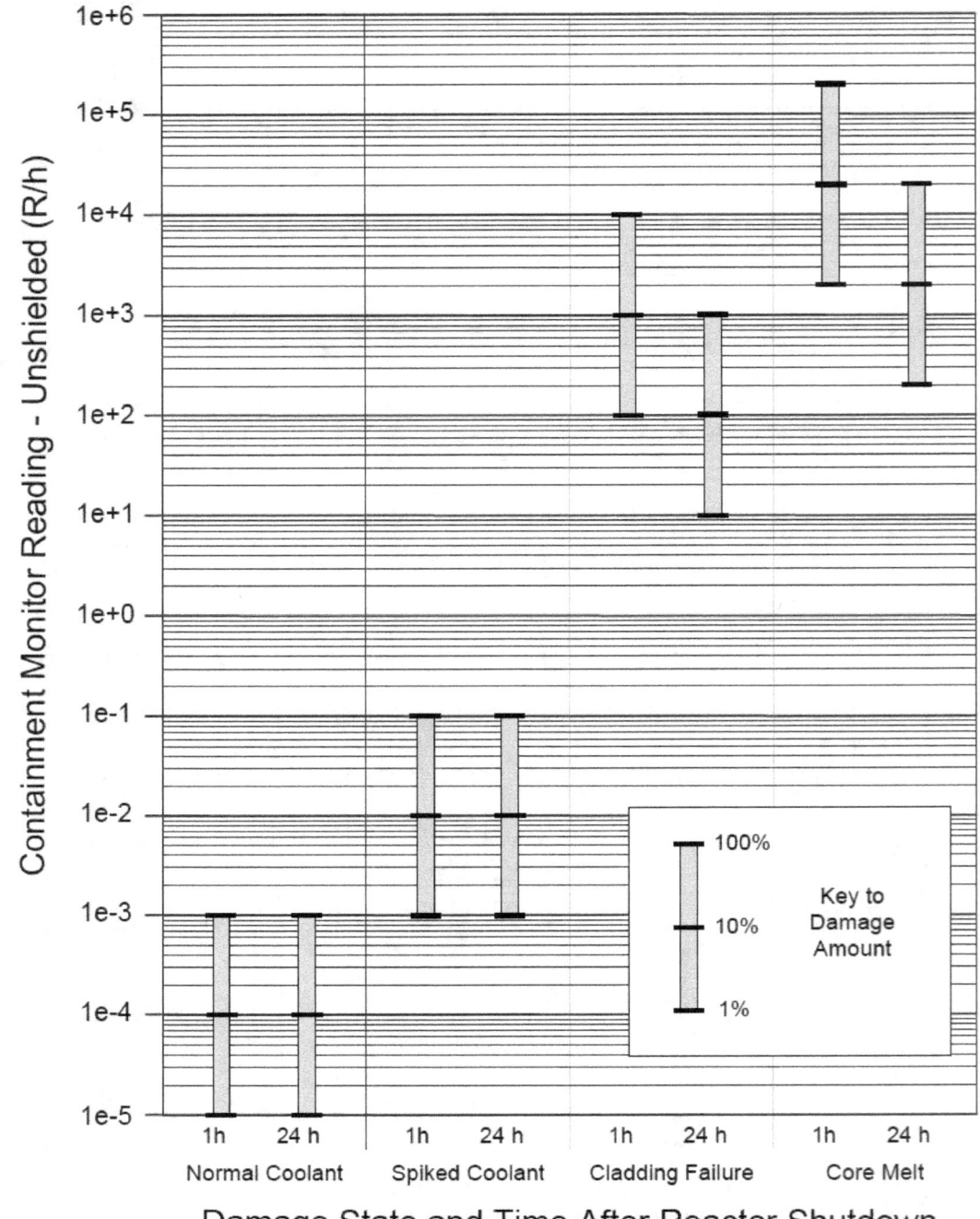

Figure 1-3 BWR Mark I and Mark II wetwell containment monitor response

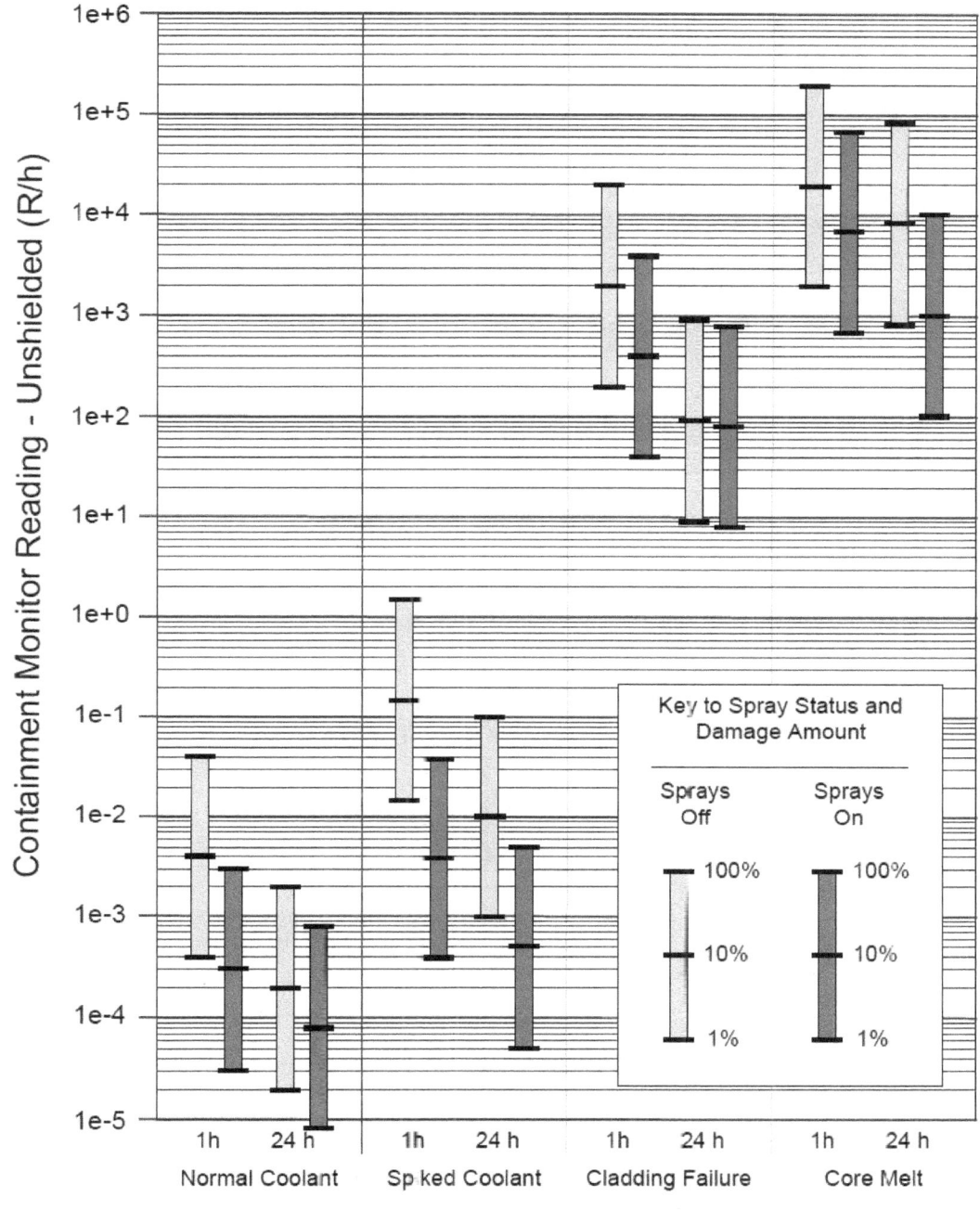

Figure 1-4   BWR Mark III drywell containment monitor response

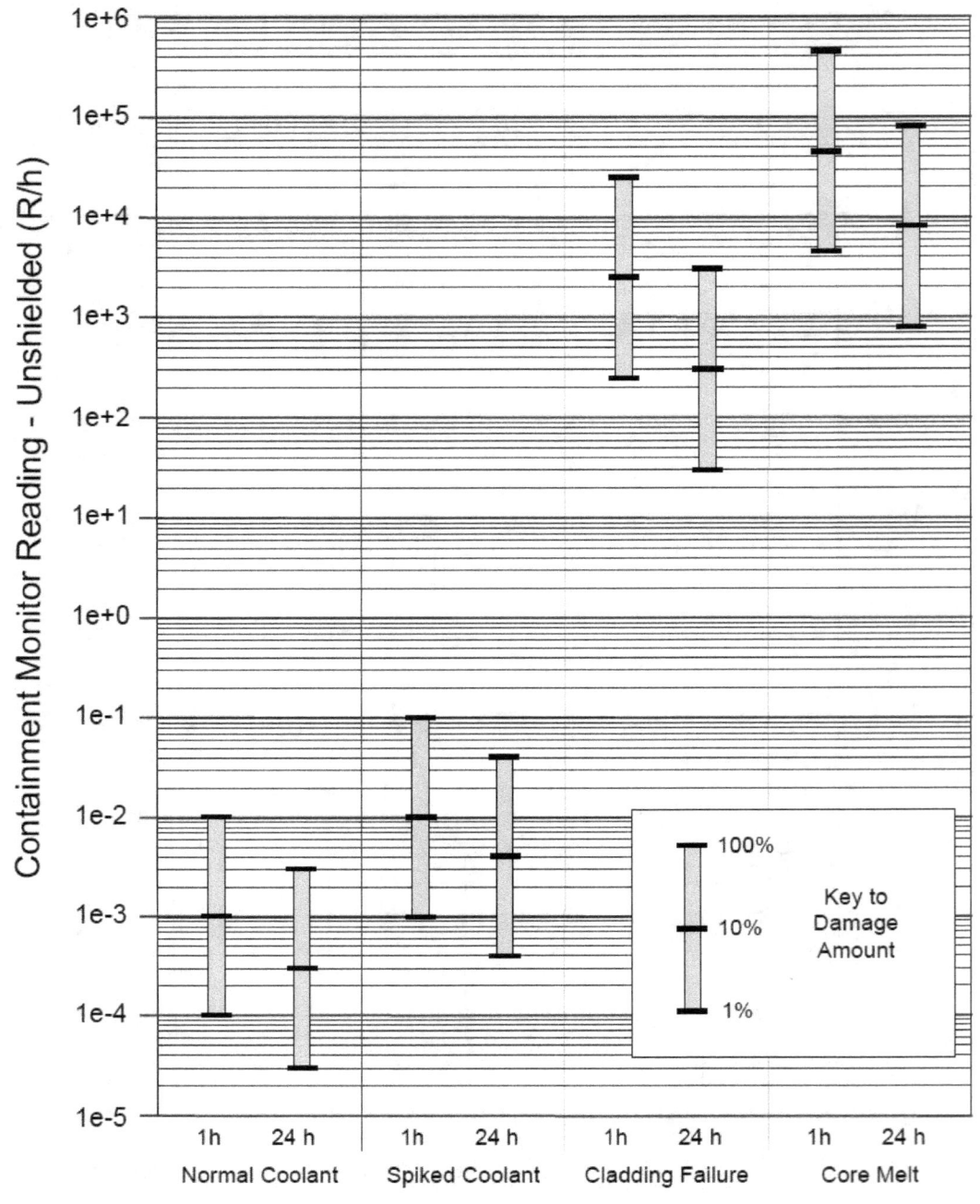

Figure 1-5   BWR Mark III wetwell containment monitor response

1.2.5 Source Term Based on Coolant Sample

The measured concentrations of radionuclides in a nuclear power plant coolant sample can be used to define the source term when the activity that is being released comes from the coolant. The user must specify coolant radionuclide concentrations by nuclide. The coolant sample can only include radionuclides available from the PWR or BWR coolant inventory (Table 1-2). The analysis of coolant activities will normally take a couple of hours to complete. Thus, the coolant sample source term option cannot generally be run early in the accident sequence.

RASCAL 4 assumes that the coolant sample represents the activity concentrations in the entire RCS at the start of the release and that the reactor is shut down. The entered concentrations are multiplied by the RCS water mass to set the total activity in the RCS. RASCAL 4 assumes that clean makeup water is being added to the primary system to maintain the water level. This makeup flow rate determines the rate at which the contaminated water is leaving the primary system. If the sample has been taken after clean makeup water has been added, the sample may represent a diluted concentration. RASCAL 4 does not compute any radioactive decay before the start of the release, but it does calculate radioactive decay after the release begins. The user must decay the activity in the coolant sample from the time that the sample is taken to the time of the beginning of release. This may involve either positive (the release starts after the sample is taken) or negative (the release starts before the sample is taken) time. The corrected activity concentration should be entered for each radionuclide rather than for the measured concentration.

Two pathways are available for coolant to leave the RCS: (1) into the SG secondary side through a tube rupture or (2) out of the containment through a break in coolant piping. This chapter later discusses the specifics of the SG tube rupture and bypass release pathways. RASCAL 4 does not model releases of coolant into the containment air volume. In general, coolant activity concentrations are low, and when these concentrations are released into containment, the release to the environment will be small.

1.2.6 Source Term Based on Containment Air Sample

RASCAL 4 can use the concentration of radionuclides measured in a containment air sample to define the activity released by a containment leakage pathway. (A containment air sample cannot be used to define the activity released in the SG tube rupture or containment bypass pathways because containment air does not exit by those pathways.)

The user enters the concentration (activity/unit volume) for each nuclide in the containment atmosphere. Multiplying the radionuclide concentrations in the containment air (activity/unit volume) times the volumetric release rate (volume/time) will equal the activity release rate (activity/time). RASCAL 4 does not compute any radioactive decay before the start of the release, but it does calculate radioactive decay after the release begins. The user must decay the activity in the containment sample from the time that the sample is taken to the time of the beginning of release. This may involve either positive (the release starts after the sample is taken) or negative (the release starts before the sample is taken) time. The corrected activity concentration should be entered for each radionuclide rather than for the measured concentration.

For PWRs, the volumes are the total containment volumes. For BWRs, the volumes are the drywell volumes plus the wetwell minimum air volumes. For BWRs, RASCAL 4 analyzes only containment air samples that are taken from the drywell.

If the containment is under pressure, the density of the containment atmosphere will be greater than the density of air at normal atmospheric pressure. RASCAL 4 does not correct for this difference. The code assumes that the measurement data have been corrected for pressure. The code assumes that the data are entered in terms of activity/unit volume at the containment pressure. If the sample results are reported for the volume at atmospheric pressure, the user should increase the reported activity to account for the higher atmospheric density in the containment before entering the sample results into the code.

1.2.7 Source Term Based on Effluent Release Rates or Concentrations

RASCAL 4 can generate a source term based on effluent measurements. The user can enter the effluent release rates (activity/unit time) by radionuclide. The air sample can only include radionuclides available from the PWR or BWR core inventory (Table 1-1). RASCAL 4 assumes that the release is directly into the atmosphere so that reduction factors (e.g., filtering) cannot be applied.

Alternatively, the user can enter the effluent concentration (activity/unit volume) by radionuclide and the flow rate (volume/unit time). RASCAL 4 computes the activity of each radionuclide released to the environment as the concentration times the volumetric release rate to the environment times the release duration. The code does not calculate any radioactive decay or ingrowth before release to the atmosphere. However, it does calculate radioactive decay after release during the transport and diffusion phase.

Up to three sets of release rates or concentrations can be entered along with their start and end times.

1.2.8 Monitored Release—Mixtures

Nuclear power plants often report effluent mixtures of radionuclides by reporting the activities of three components of the mixture: (1) noble gases, (2) iodines, and (3) particulates.

In RASCAL 4, a monitored mixture release may start before or after reactor shutdown. The measurement of the effluent release rate must occur during the release because nothing can be measured before the release occurs. The default for the time of measurement is the start of the release because the plant operators will likely note the release rate as soon as it starts.

Before shutdown, RASCAL 4 assumes that the noble gas and iodine radionuclides are in radiological equilibrium and that they are present in the same proportion as in the core inventory shown in Table 1-1. Table 1-6 shows the fraction of each noble gas isotope in the noble gas portion of the effluent. Similarly, Table 1-6 shows the fraction of each iodine in the iodine portion of the sample.

Estimating the isotopic mix for the particulate portion of the monitored mix is more difficult because of the large number of particulates that could be released. The approach that is taken in RASCAL 4 is to assume that the particles are cesium iodide (CsI). These radionuclides were selected because they are both present in the core and coolant in relatively large amounts, they are both relatively volatile and readily released from damaged fuel, and they are both biologically significant. This approach is likely to overestimate the dose from particulates; however, because releases should be filtered, the particulate release rate should be low, and an overestimate of the dose from particulates should have little practical consequence.

Starting with the radioactive isotopes of iodine and cesium in the core inventory shown in Table 1-1, the stable iodine and cesium fission products were estimated based on ORIGEN 2 estimates of total iodine and cesium production. RASCAL 4 estimates of the fraction of the particle activity associated with each isotope under the further assumption that the cesium and iodine are in proper stoichiometric proportions. Iodine is the limiting CsI constituent. Table 1-7 shows the fraction of total particle activity associated with each isotope. Note that 99.6 percent of the released activity is associated with iodine isotopes, and less than 0.4 percent is associated with cesium isotopes. Table 1-8 for CsI particles is similar to Table 1-6 in that it shows the activity for each isotope at shutdown and 1 hour later.

Table 1-6 Fraction of Total Activity for Noble Gas and Iodine Isotopes at Shutdown*

NUCLIDES	CORE AT SHUTDOWN (Ci/MWt) (from Table 1-1)	FRACTION OF ACTIVITY AT SHUTDOWN	CORE 1 HOUR AFTER SHUTDOWN (Ci/MWt)	FRACTION OF ACTIVITY 1 HOUR AFTER SHUTDOWN
Kr-83m	3,050	0.0183	2,089	0.0209
Kr-85	278	0.0017	278	0.0028
Kr-85m	6,170	0.0371	5,287	0.0528
Kr-87	12,300	0.0740	7,133	0.0713
Kr-88	17,000	0.1022	13,309	0.1330
Xe-131m	365	0.0022	364	0.0036
Xe-133	54,300	0.3261	54,002	0.5397
Xe-133m	1,720	0.0103	1,697	0.0170
Xe-135	14,200	0.0854	13,158	0.1315
Xe-135m	11,500	0.0690	793	0.0079
Xe-138	45,600	0.2737	1,942	0.0194
Total Noble Gas		1.0000		1.0000
I-131	26,700	0.1154	26,604	0.1466
I-132	38,800	0.1677	28,701	0.1582
I-133	54,200	0.2343	52,424	0.2889
I-134	59,800	0.2585	27,106	0.1494
I-135	51,800	0.2240	46,636	0.2570
Total Iodine		1.0000		1.0000

*Burnup = 38,585 MWd/MTU.

Table 1-7   Fraction of Particle Activity at Shutdown Assuming CsI Particles

NUCLIDE	CORE INVENTORY (g/MWt)	SPECIFIC ACTIVITY[a] (Ci/g)	CORE ACTIVITY[b] (Ci/MWt)	MOLES	POTENTIAL MOLES IN CsI	CsI ACTIVITY (Ci/MWt)	CsI ACTIVITY FRACTION
I-131	2.15×10^{-01}	$1.24 \times 10^{+05}$	26,700	1.64×10^{-03}	1.64×10^{-03}	$2.67 \times 10^{+04}$	1.15×10^{-01}
I-132	3.77×10^{-03}	$1.03 \times 10^{+07}$	38,800	2.85×10^{-05}	2.85×10^{-05}	$3.88 \times 10^{+04}$	1.67×10^{-01}
I-133	4.80×10^{-02}	$1.13 \times 10^{+06}$	54,200	3.61×10^{-04}	3.61×10^{-04}	$5.42 \times 10^{+04}$	2.33×10^{-01}
I-134	2.24×10^{-03}	$2.67 \times 10^{+07}$	59,800	1.67×10^{-05}	1.67×10^{-05}	$5.98 \times 10^{+04}$	2.58×10^{-01}
I-135	1.48×10^{-02}	$3.51 \times 10^{+06}$	51,800	1.09×10^{-04}	1.09×10^{-04}	$5.18 \times 10^{+04}$	2.23×10^{-01}
Stable I[c]	$5.42 \times 10^{+00}$			4.27×10^{-02}	4.27×10^{-02}		
Total I[d]	$5.70 \times 10^{+00}$		231,300	4.48×10^{-02}	4.48×10^{-02}	$2.31 \times 10^{+05}$	9.96×10^{-01}
Cs-134	$3.64 \times 10^{+00}$	$1.29 \times 10^{+03}$	4,700	2.72×10^{-02}	2.60×10^{-03}	$4.49 \times 10^{+02}$	1.93×10^{-03}
Cs-136	2.03×10^{-02}	$7.33 \times 10^{+04}$	1,490	1.49×10^{-04}	1.43×10^{-05}	$1.42 \times 10^{+02}$	6.13×10^{-04}
Cs-137*	$3.73 \times 10^{+01}$	$8.71 \times 10^{+01}$	3,250	2.72×10^{-01}	2.60×10^{-02}	$3.11 \times 10^{+02}$	1.34×10^{-03}
Stable Cs[c]	$2.25 \times 10^{+01}$			1.69×10^{-01}	1.62×10^{-02}		
Total Cs[d]	$6.35 \times 10^{+01}$		9,440	4.96×10^{-01}	4.48×10^{-02}	$9.02 \times 10^{+02}$	3.89×10^{-03}
Total I + Cs			240,740		4.48×10^{-02}	$2.32 \times 10^{+05}$	

(a) RASCAL 4 nuclide database.

(b) Table 1-1.

(c) Total fission product minus radioactive isotopes.

(d) Calculated by ORIGEN 2.2.

*Burnup = 38,585 MWd/MTU.

Table 1-8   Fraction of Particle Activity at Shutdown and 1 Hour after Shutdown Assuming CsI Particles

NUCLIDE	CsI ACTIVITY AT SHUTDOWN (Ci/MWt)	CsI ACTIVITY FRACTION AT SHUTDOWN	CsI ACTIVITY 1 HOUR AFTER SHUTDOWN (Ci/MWt)	CsI ACTIVITY FRACTION 1 HOUR AFTER SHUTDOWN
I-131	26,700	0.1150	26,604	0.1459
I-132	38,000	0.1671	28,701	0.1574
I-133	54,200	0.2334	52,424	0.2875
I-134	59,800	0.2575	27,106	0.1486
I-135	51,800	0.2231	46,636	0.2557
Total I	**231,300**	**0.9961**	**181,471**	**0.9951**
Cs-134	449	0.0019	449	0.0025
Cs-136	142	0.0006	142	0.0008
Cs-137*	311	0.0013	311	0.0017
Total Cs	**902**	**0.0039**	**902**	**0.0049**
Total I + Cs	**232,202**	**1.0000**	**182,373**	**1.0000**

The method used to estimate activity associated with the isotopes in CsI particles in RASCAL 4 is a significant change from the assumptions made in RASCAL 3.0.5. It is based on stoichiometric considerations rather than assumptions. The most noticeable change is in the activity associated with Cs-137. RASCAL 3.0.5 assumes that 50 percent of the activity associated with particles was Cs-137; the stoichiometric calculations lead to a Cs-137 activity fraction of just over 0.13 percent in RASCAL 4 and a total cesium activity fraction of less than 0.4 percent. This change will result a significant decrease in the long-term consequences of a monitored mix release in RASCAL 4 relative to RASCAL 3.0.5. The change in consequences associated with the change in iodine activity from RASCAL 3.0.5 to RASCAL 4 is more complex. Inhalation doses should decrease because the iodine-131 (I-131) activity fraction decreases from 50 percent to about 11.5 percent. Most of the activity is in the comparatively short-lived iodines (I-132, I-133, I-134, and I-135), which have much lower inhalation dose conversion factors than that of I-131. In contrast, the change in the iodine activity distribution will increase the external doses from groundshine and immersion because the short-lived iodines have dose conversion factors for these pathways that are larger than the I-131 dose factors.

RASCAL 4 has an option that allows the user to enter radioiodine release rates as "I-131 equivalent" rather than "total iodine." The code includes this option because some plants report radioiodine release rates from monitored mixtures as "I-131 equivalent." When the user selects the "I-131 equivalent" option, RASCAL 4 assumes that I-131 is the only radioiodine being released. No radiological decay correction is made because the composition of the radioiodine mixture assumed by the plant is unknown.

The plants that report radioiodine release rates from monitored mixtures as "I-131 equivalent" do it in the following manner. They first take a total iodine count rate from an iodine-absorbing media like a charcoal filter. From the total count rate, the detector efficiency, and the collection time interval, they calculate the total iodine release rate. Then, using an assumed fraction of the total release for each of the radioiodines

(I-131, I-132, I-133, I-134, and I-135), the release rate of each radioiodine nuclide is estimated as the product of total iodine release rate and the assumed fraction. Finally, they multiply the release rate of each iodine nuclide by its inhalation dose conversion factor and divide the sum of all iodine nuclides by the 50-year inhalation dose conversion factor for I-131 to obtain an "I-131 equivalent" release rate. Note that this method only calculates the inhalation dose correctly. Cloudshine and groundshine will be underestimated, but those are not the dominant dose pathways for radioiodines.

Note that the initial total activity release rate is not determined by whether the reactor is shut down because the release rates for noble gases, iodines, and particles are specified in the user input. Only the distribution of activity among the isotopes changes in the initial release rates. The last columns in Tables 1-6 and 1-8 show the isotopic release fractions at 1 hour after shutdown. Comparing these fractions with the fractions in the third columns in these tables shows the changes. As expected, the activity fractions for short-lived isotopes decrease with increasing time after shutdown; consequently, the fractions of long-lived isotopes increase. These changes have consequence implications that may be counterintuitive. For a given total activity release, increasing the time after shutdown increases the inhalation doses because of the larger fraction of significant isotopes, such as I-131. In contrast, it decreases the cloudshine because the activity fraction of isotopes, such as xenon (Xe)-135m and Xe-138, decreases.

RASCAL 4 calculates the effluent rates for individual nuclides in a monitored mixture release using the above tables in the cases described below.

1.2.8.1 Case 1—No Shutdown or Release Ends before Shutdown

If a monitored mixture release ends before shutdown, the noble gas, radioiodine, and particle constituent activities remain in the same proportions that they have in the core inventory. RASCAL calculates the activity effluent release rate for each noble gas, iodine, or particle constituent radionuclide A_i by multiplying the total noble gas, iodine, or particle effluent rate A_i by the fraction for radionuclide i in Table 1-6 or Table 1-7, as appropriate:

$$A_i = A \times F_i, \qquad\qquad (1\text{-}6)$$

where:

        A_i = the activity release rate of radionuclide i
        A = the total measured activity release rate of noble gases, radioiodines, or particle constituents, as appropriate
        F_i = the fraction of nuclide i in the mixture from Table 1-6 or Table 1-7, as appropriate

In this case, the isotopic release rates remain constant for the duration of the release.

1.2.8.2 Case 2—Release Starts before Reactor Shutdown and Ends after Reactor Shutdown, and Measurement Is Made before or at Reactor Shutdown

If the sample measurement is made at or before reactor shutdown, each noble gas and iodine nuclide is present in the monitored sample in the fractions shown in Tables 1-6 and 1-7. Thus, the release rate for each nuclide for any time step before shutdown is simply the effluent rate multiplied by the appropriate fraction in Table 1-6 or Table 1-7, as shown in Equation 1-6.

The isotopic release rates at or before shutdown from Equation 1-6 are not corrected for radioactive decay. Thus, the total activity release rate is constant. For time steps after shutdown, the total isotopic release rates decrease because of radiological decay.

$$A_i(t) = A_{0i}\exp[-\lambda_t(t - t_0)],\qquad(1\text{-}7)$$

where:

   $A_i(t)$ = the decay-corrected activity release rate for nuclide i at time t
   A_0i = the activity release rate of nuclide i at time of shutdown t_0
   λ_i = the radiological decay constant of nuclide i

Thus, the total activity release rate does not remain constant; instead, it decreases with time. (The ingrowth of daughters from the decay of the noble gases is not included in the release because they would be filtered out before release.)

1.2.8.3 Case 3—Release Starts before Reactor Shutdown and Ends after Reactor Shutdown, and Measurement Is Made after Reactor Shutdown

RASCAL 4 performs this calculation in three steps. First, it determines an unnormalized isotopic fraction for each radionuclide in the mixture sample $F_i(t)$ for the sample time t_s by applying a decay correction to the fractions in Tables 1-6 and 1-7, as follows:

$$\bar{F}_i(t_s) = F_{0i}\exp[-\lambda_t(t_s - t_0)],\qquad(1\text{-}8)$$

where:

   $F_i(t_s)$ = the unnormalized decay-corrected isotopic fraction for nuclide i at sample time t_s
   F_{0i} = the isotopic fraction of nuclide i at time of shutdown time t_0 from Table 1-6 or Table 1-7
   λ_i = the radiological decay constant of nuclide i

Next, the code calculates the isotopic release rate for each radionuclide in the sample $A_i(t_s)$ by normalizing the fractions and multiplying by the effluent release rate for that radionuclide group

$$A_i(t_s) = A(t_s)\left[\frac{F_i}{\sum F_{si}}\right],\qquad(1\text{-}9)$$

where:

   $A_i(t_s)$ = the isotopic release rate of nuclide i at sample time t_s (after shutdown)
   $A(t_s)$ = the total noble gas or radioiodine activity release rate at the sampling time t_s
   ΣF_{si} = the sum of the fractions of the activities for the radionuclide group at the sampling time (to normalize the fractions for each nuclide)

The last step is to calculate the initial isotopic release fractions by correcting the sample isotopic release fractions back to the start of release to account for radiological decay:

$$A_i(t) = A_i(t_s)\exp[-\lambda_i(t - t_s)],\qquad(1\text{-}10)$$

where:

   $A_i(t)$ = the decay-corrected activity effluent release rate for nuclide i at time t
   $A_i(t_s)$ = the activity effluent release rate of nuclide i at sample time t_s
   λ_i = the radiological decay constant of nuclide i

Note that if the time t is earlier than the sample time t_s, the exponent in the equation will be positive and the radiological decay correction will increase the isotopic release fractions. They will no longer be normalized to a value of 1.0. In addition, the activity effluent release rates will be greater than they were at the sample time.

RASCAL 4 will calculate the initial isotopic release rates as if the release started at reactor shutdown because the start of release is at or before reactor shutdown. In this case, the isotopic release rates will be constant from the start of release to the time of reactor shutdown. After shutdown, the isotopic release rates will decrease from radioactive decay.

1.2.8.4 Case 4—Release Starts and Ends after Reactor Shutdown

The calculations in this case are similar to those in Case 3 above. RASCAL 4 initially calculates the isotopic release rates for the time of measurement and then adjusts them to the start of release. If the measurement occurred after the start of release, the initial total activity releases for noble gases, iodines, and particles will be greater than the activity releases specified by the user. However, the isotopic and total activity release rates will decrease by radioactive decay as the release progresses. At the measurement time, the release rates will be those specified by the user.

1.3 Release Pathways

After the RASCAL 4 user has selected a source term type and has entered the needed data for that source term type, he or she must select a release pathway to the environment. The release pathways that are available for selection will depend on the reactor type (PWR or BWR) and the source term type that the user selected.

For PWRs, four potential release pathways exist: (1) containment leakage, (2) containment bypass, (3) SG tube ruptures, and (4) direct to atmosphere. Table 1-9 shows the release pathways available for those source term types.

For BWRs, four potential release pathways also exist: (1) leakage from the drywell through the wetwell, (2) leakage through the drywell wall, (3) containment bypass, and (4) directly to the atmosphere. Table 1-10 shows the release pathways available for each source term type.

Table 1-9   PWR Release Pathways Available for Each Source Term Type

SOURCE TERM TYPE	RELEASE PATHWAY			
	Containment Leakage	Containment Bypass	SG Tube Rupture	Directly to the Atmosphere
Time core is uncovered	X	X	X	
Final core damage endpoint, with spiked coolant release		X	X	
Final core damage endpoint, with cladding damage	X	X	X	
Containment monitor readings and containment air sample	X			
Coolant sample		X	X	
Effluent releases (rates, concentrations, and mixtures)				X

Table 1-10   BWR Release Pathways Available for Each Source Term Type

SOURCE TERM TYPE	RELEASE PATHWAY			
	Leakage from the Drywell through the Wetwell	Leakage from the Drywell through the Drywell Wall	Bypass Containment	Directly to the Atmosphere
Time core is uncovered	X	X	X	
Final core damage endpoint, with spiked coolant release			X	
Final core damage endpoint, with cladding damage	X	X	X	
Containment monitor readings and containment air sample	X	X		
Coolant sample			X	
Effluent releases (rates, concentrations, and mixtures)				X

1.4 Release Pathway Models and Reduction Mechanisms

Each of the pathways listed in the previous section, except the direct release to the atmosphere, will have its own characteristic potential reduction mechanisms. The reduction factors that RASCAL 4 uses are described below.

RASCAL 4 assumes that all the reduction factors operate on all radionuclides except noble gases. None of the reduction factors reduce the activity of the noble gas release to the environment. The code assumes that all nuclides subject to a given reduction mechanism have the same reduction factor. It treats radioiodines the same as all other nonnoble gas nuclides. Table 1-11 lists the reduction factor multipliers, and the sections below describe them in more detail.

Table 1-11   Summary of Nuclear Power Plant Reduction Factor Multipliers*

REDUCTION MECHANISM OR CAUSE	REDUCTION FACTOR MULTIPLIER
Containment sprays (Reference: Figure 5 in NUREG/CR-4722, "Source Term Estimation Using MENU-TACT," issued 1987 (Sjoreen et al., 1987))	First 0.25 hour: exp(-12t) After 0.25 hour: exp(-6t)
Containment natural processes during hold-up (Reference: Appendix B to NUREG-1150, "Severe Accident Risks: An Assessment for Five U.S. Nuclear Power Plants," issued December 1990 (NRC, 1990))	First 1.75 hour: exp(-1.2t) 1.75 to 2.25 hours: exp(-0.64t) After 2.25 hours: exp(-0.15t)
PWR Ice condenser—no fans or recirculation	0.5
PWR Ice condenser—1 hour or more recirculation	0.25
BWR release pathway from the drywell through the wetwell with subcooled pool water	0.01
BWR release pathway from the drywell through the wetwell with saturated pool water	0.05
Plate-out for containment bypass pathway	0.4
SG tube rupture—partitioned (break underwater)	Partitioning factor (steam concentration as fraction of SG water concentration) 0.02
SG tube rupture—not partitioned (break above water level)	Partitioning factor (steam concentration as fraction of SG water concentration) 0.5
SG tube rupture—condenser off-gas release	0.05
SG tube rupture—safety relief valve release	1
Filters	0.01
Lower limit on reduction multiplier (except for filters)	0.001
Lower limit on reduction multiplier for containment sprays (Reference: Figure 5 in NUREG/CR-4722)	0.03

*Reference: NUREG-1228, except as noted for some specific table lines.

1.4.1 Containment Leakage in Pressurized-Water Reactors

While radionuclides are held up in the containment atmosphere, they are subject to removal from the atmosphere by water sprays and by natural processes that cause deposition on containment surfaces. If containment sprays are operating, they rapidly reduce the concentrations of all radionuclides except for noble gases. If the sprays are not operating, the natural processes, such as gravitational settling and plate-out on containment surfaces by turbulent impaction, gradually reduce the airborne concentrations of particulates and reactive gases.

The reduction factors RF for sprays and for natural processes during holdup without sprays are modeled as exponential functions of time t, as follows:

$$RF = e^{-\lambda t}, \tag{1-11}$$

where λ is a reduction constant for sprays or natural processes.

Both sprays and natural processes have multiple values for λ. The removal rate is larger at early times and slower at later times. Sprays and natural processes can remove particulates more readily initially and then more slowly after the removal of the readily removable particles.

Because the user can enter release and reduction data that change with time, the sprays can be turned on and off several times. The initial spray λ_I applies to (1) all the activity in containment the first time that the sprays are turned on and (2) all the activity that enters the containment the first time that the sprays are active. If the sprays are turned off and then turned back on, RASCAL 4 uses only the continuing λ_C. The initial λ_I for holdup applies only if the sprays were never turned on; otherwise, the code uses the continuing λ_C.

RASCAL 4 nuclear power plant source term calculations include a maximum effectiveness for sprays and a maximum effectiveness for all reduction, excluding filters (Table 1-11). For each, RASCAL 4 compares the appropriate reduction factor or product of reduction factors computed at each time step to the maximum and does not allow either one to surpass it.

For PWRs with ice condenser containments, the user can take additional reductions because of the interaction of the containment air with the ice. If the fans are recirculating the containment air through the ice beds for at least 1 hour, RASCAL 4 reduces the activity entering the containment using a reduction factor RF_i multiplier of 0.25. If the fans are not operating, the reduction factor RF_i multiplier is 0.5. After the ice beds are exhausted, the reduction factor RF_i multiplier is 1.

1.4.2 Containment Leakage in Boiling-Water Reactors

For BWRs, the model for reduction of radionuclides in the drywell air by sprays or natural processes is the same as that for PWRs, as described above. However, the user can apply an additional reduction mechanism if the release from the drywell is through the wetwell water.

If the release is through the wetwell water and if the water is subcooled (below the boiling point), RASCAL 4 applies an additional reduction factor RF_i multiplier of 0.01 to all nuclides except noble gases. If the wetwell water is saturated (boiling), the reduction factor RF_i multiplier is 0.05.

1.4.3 Containment Bypass

Containment bypass is a coolant release from the RCS to an auxiliary building or directly to the environment without passing through the containment atmosphere. The containment bypass release model and the reduction mechanisms are the same for PWRs and BWRs. Therefore, this section applies equally to both.

For the bypass model, RASCAL 4 first calculates the initial concentration of each radionuclide in the coolant. If the user selects the coolant source term type, the initial coolant concentration of each radionuclide is entered directly. If the user selects the ultimate core damage state source term type, the code assumes that the radionuclide activity by nuclide (calculated as described in Section 1.2.3) enters the primary coolant system. The initial concentration of each radionuclide is the activity entering the primary coolant system divided by the total coolant volume. If the user selects the time core is uncovered source term type, the initial coolant concentration is the activity released from the core during the first 15-minute time step divided by the total coolant volume.

The coolant concentrations are then multiplied by a reduction factor for plate-out. The plate-out multiplier for containment bypass is 0.4, which is taken from NUREG-1228. The plate-out mechanism is plate-out within the RCS.

The user then enters the coolant escape rate in terms of volume per unit time. Generally, the user can estimate the escape rate based on the makeup needed to maintain water levels. RASCAL 4 then calculates the activity that escapes the primary coolant system during the time step by multiplying the radionuclide concentration in the coolant times the volume that escapes during the first time step.

For subsequent time steps, RASCAL 4 reduces the concentration in the coolant to account for the activity that has escaped from the primary system. If the user selects the time core is uncovered source term type, new activity enters the coolant system during each time step, as described in Section 1.2.2. Thus, the coolant concentrations are being augmented during each time step by the entrance of new radioactive material into the coolant.

RASCAL 4 also calculates radioactive decay and ingrowth during each time step.

RASCAL 4 does not calculate any holdup or plate-out in any secondary structure, such as the auxiliary building. However, filters can reduce a release to the environment, if applicable.

1.4.4 Steam Generator Tube Ruptures in Pressurized-Water Reactors

RASCAL 4 calculates the activity concentration in the primary coolant system the same manner in which it calculates the concentration for containment bypass, as described in Section 1.4.3 above. The code also calculates activity escaping the primary coolant system and entering the SG by the same method used for bypass accidents except that no reduction factor for plate-out is used.

As with the bypass release path, the RASCAL 4 user specifies the flow rate from the primary coolant system to the secondary system, which can perhaps be estimated from the makeup flow needed to maintain the water level in the primary system. The default flow rate into the SG is 500 gallons per minute, which is considered equivalent to the rupture of one tube in one of the SGs.

For U-tube steam generators, the RASCAL 4 user specifies whether the tube rupture is above or below the water level in the SG. For once-through steam generators, the code always assumes that rupture is above the water level.

If the break is below the water level, RASCAL 4 assumes that the activity entering the SG is evenly mixed in the SG water. The initial activity concentration (curies per pound) in the SG is the activity that entered the SG during the first time step divided by the weight of the SG water. The default mass of water in a SG is 42,184 kg (93,000 pounds).

RASCAL 4 assumes that the activity concentration for nonnoble gases in the steam that exits the SG is the concentration in the SG water times a partition factor. If the rupture is below the water level, the partition factor is 50. In other words, the code assumes that the concentration of a nonnoble gas radionuclide in the steam is one-fiftieth (0.02) of the concentration in the SG water. If the break is above the water level, the partition factor is 2. In other words, the concentration in the steam is one-half of the concentration in the water.

Note that partition factors are holdup factors, not removal factors. The partition factors slow the release of radionuclides from the SG, but they do not prevent it. As long as the SG is not isolated, the steam will continue to remove radionuclides from the SG water. The removal rate for the steam is the concentration in the steam times the flow rate of steam. The default for the steaming rate in an SG is 34,019 kg per hour (75,000 pounds per hour). This value is the amount of steam needed to remove decay heat soon after shutdown. The user can change this value with time if better information is available.

At each time step, RASCAL 4 recalculates the concentration of radionuclides in the SG water by subtracting the activity removed in the steam during the previous time step and by adding any new activity entering from the primary coolant system through the rupture.

Two paths by which the radionuclides in the steam can escape to the environment exist. The first is the safety relief valve, and the other is the condenser off-gas exhaust (or steam-jet air ejector in some plants).

RASCAL 4 assumes that there is no removal of radionuclides as the steam exits through the safety release valve. If the exit is through the condenser off-gas exhaust, RASCAL 4 assumes the removal of 95 percent of the nonnoble gas radionuclides (multiplying the activity by 0.05).

1.4.5 Boiling-Water Reactor Release through the Standby Gas Treatment System

The SBGTS is available on BWRs. It provides a high-volume draw of air that is passed through filters and released from the stack. RASCAL 4 assumes that the SBGTS can draw from the reactor building (secondary containment) of Mark I and Mark II containments, the annulus of the Mark III containment, and adjacent building volumes into which material might be released (such as the turbine building). The assumption is that the flow rate of the SBGTS is high enough to keep up with any leak rate from the primary containment. Therefore, the user interface does not gather leak or flow rates from the SBGTS.

When the user specifies that the release is through the SBGTS, the release height is fixed at the stack height contained in the RASCAL 4 facility database. In addition, the release is always filtered. If the release is not through the SBGTS, RASCAL 4 assumes that the release is directly from the reactor building to the atmosphere or through some other rapid, unfiltered release pathway.

1.5 Leakage Fractions

Four methods for specifying leakage fractions for release to the environment are available in RASCAL 4: (1) specifying the percentage of activity present that is released per unit time, (2) specifying a containment pressure and hole size, (3) specifying a coolant flow rate (volume or mass per unit time), and (4) specifying a "direct" release with all activity released during the selected release duration. Table 1-12 shows the methods used for specifying release rates available for each release pathway.

Table 1-12 Methods for Specifying Release Rate for Each Release Pathway

RELEASE PATHWAY	METHOD FOR SPECIFYING RELEASE RATE
Containment leakage	Percentage of containment volume per time or containment pressure and hole size
Containment bypass	Coolant flow rate (makeup flow to maintain RCS water level)
SG tube rupture	Coolant flow rate plus steaming rate
Monitored effluent releases	Directly to the atmosphere

1.5.1 Percent Volume per Time

This release rate method releases the activity in a fixed fraction of the containment or confinement volume per unit time.

RASCAL 4 uses the leakage fraction *LF* to calculate the fraction of the radionuclide inventory in the containment atmosphere that is released to the environment during each 15-minute time step. At each time step, the code adjusts the radionuclide inventory in the containment atmosphere to account for (1) radiological decay and ingrowth, (2) additions to the containment atmosphere radionuclide inventory if core damage is still occurring, (3) removal of radionuclides from the containment atmosphere by sprays or plate-out, and (4) removal from the containment atmosphere by release to the environment.

Consider the case in which the release rate is specified to be 100 percent, which corresponds to total containment failure. This rate is equal to 25 percent per 15-minute time step. During the first time step, release of 25 percent of the activity in the containment will occur. For the second time step, RASCAL 4 will reduce the activity remaining in the containment by subtracting the activity that escaped during the first time step. Then the code will calculate any applicable reduction factors, such as removal by containment sprays or plate-out. Then it will apply the 25-percent leak to the remaining activity in the containment. Because removal of only 25 percent of the material in the containment can be done at each time step, some activity will remain in the containment after 1 hour even at a leak rate of 100 percent per hour.

1.5.2 Leak Rate Based on Containment Pressure and Hole Size

RASCAL 4 can calculate the leak rate through a hole in the containment if the hole size and containment pressure are known. RASCAL 4 uses a simplified form of Equation 6-39 from the *Applied Fluid Dynamics Handbook* (Blevins, 1984) for incompressible flow through a thin square-edged orifice. The code assumes that the hole is the orifice. The mass flow rate out of containment *MFR(k)* during time step *k* is as:

$$MFR(k) = C\left(\frac{\pi D^2}{4}\right)\sqrt{2\rho(P_1(k) - P_2)g},$$ (1-12)

where:

$C = 0.63$, an experimentally measured dimensionless discharge coefficient that rarely varies outside the range of $0.59 < C < 0.65$
D = hole diameter in inches
ρ = density of containment atmosphere in pounds per cubic inch

$P_1(k)$ = pressure in the containment during time step k in pounds per square inch

P_2 = atmospheric pressure in pounds per square inch

g = acceleration of gravity in inches per second per second to convert between pounds and a mass unit

The leakage fraction from containment to the atmosphere during step k $(LF(k))$ is:

$$LF(k) = \frac{MFR(k)t}{\rho V_c},$$ (1-13)

where:

t = duration of time step k in seconds

V_c = the containment volume

If the containment pressure is less than the atmospheric pressure, the leak rate is zero. The code does not compute the change in containment pressure, but the user can enter changing containment pressures as the assessment proceeds.

1.5.3 Coolant Flow Rate

Containment bypass accidents are accidents in which coolant is released without going through the containment. When the coolant escapes, it is no longer pressurized. At atmospheric pressure, the coolant will flash into steam, and the radionuclides in the coolant will become airborne. The coolant mass flow rate times the radionuclide concentration will give the radionuclide release rate. Alternatively, the coolant mass flow rate divided by the total coolant volume will give the leakage fraction for the radionuclides in the coolant.

Normally, measurement of the coolant mass flow rate directly cannot be done. However, determining the makeup flow needed to maintain pressure or water levels in the RCS can usually be done. RASCAL 4 can use this makeup flow as an estimate of the mass flow rate for escaping coolant.

For SG tube rupture accidents, RASCAL 4 can use the makeup flow rate as an estimate of the coolant mass flow rate from the RCS to the SG. The steaming mass flow rate in the SG is then a measure of the rate at which water is being removed from the SG. The radionuclide concentration in the steam will be the concentration in the SG water times the appropriate partitioning factor. The concentration in the steam times the steaming mass flow rate equals the escape rate from the SG.

1.5.4 Direct Release to Atmosphere

RASCAL 4 uses the direct release to the atmosphere pathway with the three monitored effluent release source term types: (1) activity release rate by nuclide, (2) activity release concentration by nuclide and flow rate, and (3) monitored mixtures release rate. The code assumes that these releases are measured after the action of any removal or reduction processes and that they represent the actual release rate to the atmosphere. Therefore, applying reduction mechanisms to the releases cannot be done.

The user sets a start and a stop time for the release. If the source term type is activity release rate by nuclide or activity release concentration by nuclide and flow rate, RASCAL 4 assumes that the activity release rate and the composition of the effluent are constant over the interval. If the user selects the monitored mixture source term type, the release rate and composition of the effluent changes with time to account for radiological decay, as described in Section 1.2.8.

1.6 Decay Calculations in the Source Term

Most of the reactor source terms require the calculation of the decay of radionuclides and ingrowth of progeny. Table 1-13 indicates which source term types calculate radioactive decay before the start of release and which calculate the decay during the release. These calculations are done using the procedures described in detail in Appendix A. RASCAL 4 uses the Bateman equations (Bateman, 1910) and decay chain information in Appendix A to calculate decay and ingrowth from the time of reactor shutdown until the release to the environment begins. The code also uses them to calculate decay and ingrowth factors for each radionuclide for a 5-minute period. The code uses these factors to decay the source term during the release.

Table 1-13   Source Term Types Calculating Decay Before and During Release

SOURCE TERM	DECAY BEFORE START OF RELEASE	DECAY DURING RELEASE
Time core uncovered	Yes	Yes
Core damage state	Yes	Yes
Containment radiation monitor	Yes	Yes
Coolant sample	No	Yes
Air sample	No	Yes
Nuclide release rate	No	No
Nuclide concentration	No	No
Monitored mix	Yes[1]	Yes[2]

(1) Activity is corrected from the time of measurement to the shutdown time (+ or or -), simple decay without daughters.

(2) Monitored mix is adjusted for decay; no other factors that might change the release rate are considered.

1.7 Verification of the Source Term Calculations

The source term calculation module for RASCAL 4 has undergone numerous verification procedures Section 1.7.1 describes numerical verification of the code. Section 1.7.2 describes a practical evaluation of the source term calculations based on modeling of the Fukushima Daiichi accident.

1.7.1 Numerical Verification

Development of RASCAL 4 included many modifications of the code used to calculate source terms. The code used for source term calculations generates a large file that includes intermediate computational results. As a result, the verification of the source term calculations has included extensive evaluation of intermediate computational results and the final source terms. Verification of the calculations was done primarily by developing spreadsheets that represent the source term models described in this document and by then comparing the results of the spreadsheet calculations with the results of the RASCAL 4 source term calculations.

A spreadsheet for each source term type contains the equations representing the methods and models described in this report. As such, the spreadsheets represent how the calculations should be correctly done. The spreadsheet for each source term type models the combination of release pathways, reduction mechanisms, and other operating conditions that are appropriate for the source term type. For example, the spreadsheet for PWR accidents based on the "time core is uncovered" involving releases through containment models sprays and filters. Similarly, spreadsheets model other PWR and BWR accidents and release paths, and they model releases based on air and coolant samples and monitoring data.

Each spreadsheet tracks the releases of several nuclides selected to evaluate various aspects of the calculations. The spreadsheets track noble gases because no reduction mechanism should operate on noble gases. The spreadsheets track iodines and cesiums because of their importance, and they track short- and long-lived isotopes to evaluate the calculation of decay and progeny ingrowth.

The basic radioactive decay algorithms used in RASCAL 4 source term calculations are also used in the transport and dispersion calculations and in FMDose. Appendix A describes the algorithms in detail. Tests of the algorithms compare parent decay and daughter ingrowth and decay for a number of isotopes to the results of hand calculations. They also include comparisons of the half-lives of nuclides in the time-dependent source terms generated for the transport, dispersion, and dose calculations with the known half-lives of nuclides. Half-lives calculated from the source term output are consistent with half-lives in the RASCAL 4 decay chain definition file. The results of the spreadsheet and RASCAL 4 source term calculations generally agree to 2 significant figures; in many cases, they agree to 3 significant figures. Round-off errors are the primary source of the differences between the spreadsheet source terms and the RASCAL 4 source terms.

Chapter 7 includes additional verification and validation of the decay and ingrowth algorithms where FMDose computational results are compared to computational results from Turbo FRMAC 2011 (SNL 2011).

1.7.2 Fukushima Daiichi Nuclear Accident

A magnitude 9.0 earthquake occurred off the eastern coast of Japan at 1446 local time on March 11, 2011. The earthquake resulted in shutdown of three operating BWR nuclear reactors at the Fukushima Daiichi site and loss of all offsite power. Emergency generators started and were operating as expected when a tsunami struck the site about 45 minutes later. The tsunami disabled the generators and changed the event classification from loss of offsite power to station blackout. Subsequently, loss of cooling resulted in damage to the reactor cores of Fukushima, Units 1, 2, and 3, and in the release of radioactive material to the environment. Table 1-14 lists several estimates of the I-131 and Cs-137 activity released to the environment. These estimates are consistent and can be used to provide insight into the adequacy of the RASCAL 4 reactor source term calculation for reactor accidents like those that occurred at Fukushima Daiichi.

The remainder of this section presents the results of a series of RASCAL 4.2 model runs related to the Fukushima accident. The runs are for existing U.S. reactors that are similar to Fukushima Units 1 and 2. The Duane Arnold Energy Center (Duane Arnold) reactor was used as a surrogate for Fukushima, Unit 1, and the Cooper Nuclear Station (Cooper) reactor was used as a surrogate for Fukushima, Units 2 and 3.

The first set of RASCAL 4 model runs uses default reactor parameters and the NUREG-1465 release sequence (Soffer, 1995) to evaluate the potential I-131 and Cs-137 releases for the first 48 hours that are associated with various BWR release pathways. Table 1-15 presents the estimates of potential I-131 and Cs-137 releases as a function of the release path from the reactor core to the environment. The delay times in the table are the approximate delay times between reactor shutdown and the beginning of core uncover for the Fukushima reactors. The 48-hour release period for each reactor starts as the core begins to uncover.

Table 1-14   Total Radionuclide Release Estimates from Fukushima, Units 1, 2, and 3

AGENCY	I-131	Cs-137
Nuclear Safety Commission of Japan (August 23, 2011)[a]	1.3×10^{17} Bq	1.1×10^{16} Bq
Nuclear and Industrial Safety Agency[a]	1.6×10^{17} Bq	1.5×10^{16} Bq
Japan Nuclear Energy Safety Organization[a]	1.3×10^{17} Bq	6.1×10^{15} Bq
Inoue (2012)	<3% of the core inventory	<2% of the core inventory
Chang et al. (2012)[b]	3% to 10% of the core inventory	2% to 3% of the core inventory

(a) As reported by Inoue (2012).

(b) Based on a MELCOR analysis of an unmitigated station blackout for Peach Bottom Atomic Power Station, (NUREG--1935, 'State-of-the-Art Reactor Consequence Analysis (SOARCA) Report," Draft report for comment.

Table 1-15 Surrogate Reactor Release Estimates

FUKUSHIMA REACTOR	DELAY (hours)[a]	RELEASE PATH					TOTAL RELEASE (Bq)	
		WETWELL	DRYWELL	BYPASS	SBGTS	DIRECT	I-131	Cs-137
Unit 1 surrogate (Duane Arnold)	4.5	X			X		$1.0\times10^{+11}$	$1.1\times10^{+10}$
	4.5	X				X	$1.0\times10^{+13}$	$1.1\times10^{+12}$
	4.5		X			X	$1.0\times10^{+15}$	$1.1\times10^{+14}$
	4.5			X			$4.4\times10^{+17}$	$4.3\times10^{+16}$
Unit 2 surrogate (Cooper)	73.5	X			X		$1.0\times10^{+11}$	$1.4\times10^{+10}$
	73.5	X				X	$1.0\times10^{+13}$	$1.4\times10^{+12}$
	73.5		X			X	$1.2\times10^{+15}$	$1.4\times10^{+14}$
	73.5			X			$4.5\times10^{+17}$	$5.4\times10^{+16}$
Unit 3 surrogate (Cooper)	37.5	X			X		$1.2\times10^{+11}$	$1.4\times10^{+10}$
	37.5	X				X	$1.2\times10^{+13}$	$1.4\times10^{+12}$
	37.5		X			X	$1.2\times10^{+15}$	$1.4\times10^{+14}$
	37.5			X			$5.1\times10^{+17}$	$5.4\times10^{+16}$
Total Release		X			X		$3.2\times10^{+11}$	$3.9\times10^{+10}$
		X				X	$3.2\times10^{+13}$	$3.9\times10^{+12}$
			X			X	$3.2\times10^{+15}$	$3.9\times10^{+14}$
				X			$1.4\times10^{+18}$	$1.5\times10^{+17}$

(a) Institute of Nuclear Power Operations (INPO), 2011

The combined RASCAL release estimates for I-131 and Cs-137 range over about 7 orders of magnitude depending on the release path. A comparison of these estimates to the release estimates in Table 1-14 suggests that the Fukushima reactor releases were through the drywell pathway. Information presented by Matsunaga (2012) supports this assumption.

Given the results of the screening effort shown above, RASCAL 4 modeled the Fukushima reactor releases using a drywell release path without sprays in the second set of runs. Table 1-16 presents the parameter values for the Fukushima reactors used for detailed modeling of the Fukushima reactor releases. Each reactor was modeled separately, and the results were combined to obtain the total for the accident.

Table 1-16   Fukushima Reactor Parameters

	UNIT 1	UNIT 2	UNIT 3	TOTAL
Uranium (metric ton)[a]	69	94	94	
Power (MWt)[a]	1,380	2,381	2,381	
Last startup[a]	9/27/2010	9/23/2010	11/18/2010	
Burnup (MWd/MTU)[b]	23,700	30,000	28,500	
I-131 inventory (Bq)[c]	$1.36\times10^{+18}$	$2.35\times10^{+18}$	$2.35\times10^{+18}$	$6.07\times10^{+18}$
Cs-137 inventory (Bq)[d]	$1.02\times10^{+17}$	$2.23\times10^{+17}$	$2.11\times10^{+17}$	$5.36\times10^{+17}$
(a) INPO (2011). (b) Based on an 18-month refueling cycle with a 30-day outage with one-third of the core replaced each cycle. (c) Based on Table 1-1 and power level. (d) Based on Table 1-1 and power level with Equation 1-1 correction for burnup.				

The RASCAL 4 core uncovered time for each reactor was set to the time that the reactor water level dropped below the top of the active fuel. The cores were not recovered. RASCAL 4 only allows the core uncovered elapsed time to be 12 hours if the core is recovered. The cores of both Fukushima, Units 1 and 3, were uncovered for more than 12 hours (INPO, 2011).

The containment leak rate for each reactor was adjusted to follow the sequence of events set out in the INPO report. The RASCAL 4 default leak rate is 0.5 percent per day. This leak rate was changed from 0.5 percent per day to 1 percent per hour at the beginning of core damage. It was changed to 25 percent per hour for 1 hour for containment venting and to 50 percent per hour for 1 hour following the Unit 1 and Unit 3 hydrogen (H_2) explosions. Following containment venting and H_2 explosions, the leak rate was returned to 1 percent per hour.

Table 1-17 presents the results of this more detailed modeling of the Fukushima releases. The magnitudes of the I-131 and Cs-137 releases shown in the table are consistent with, but somewhat higher than, other release estimates. The differences in release magnitudes are not significant because the selection of parameter values, particularly for the drywell leak rates, was not based on knowledge of the true plant conditions. One possible explanation for the differences is that RASCAL 4 does not model the secondary containment of BWRs; activity released from the primary containment goes directly to the environment.

Table 1-17   RASCAL 4.2 Estimates of Fukushima I-131 and Cs-137 Releases

	UNIT 1	UNIT 2	UNIT 3	TOTAL
I-131 inventory (Bq)	$1.36\times10^{+18}$	$2.35\times10^{+18}$	$2.35\times10^{+18}$	6.07×10^{-18}
I-131 release (Bq)	$4.85\times10^{+16}$	$9.4\times10^{+16}$	$5.6\times10^{+16}$	$2.0\times10^{+17}$
I-131 release fraction	3.6%	4.0%	2.4%	3.3%
Cs-137 inventory (Bq)	$1.02\times10^{+17}$	$2.23\times10^{+17}$	$2.11\times10^{+17}$	5.36×10^{-17}
Cs-137 release (Bq)	$4.10\times10^{+15}$	$1.13\times10^{+16}$	$6.28\times10^{+15}$	2.17×10^{-16}
Cs-137 release fraction	4.0%	5.1%	3.0%	4.1%

Figure 1-6 shows the cumulative I-131 and Cs-137 releases for the Fukushima event as a function of time after the initiating event, the March 11, 2011, earthquake. The step increases in release correspond to the increases in release rates associated with containment venting and H₂ explosions.

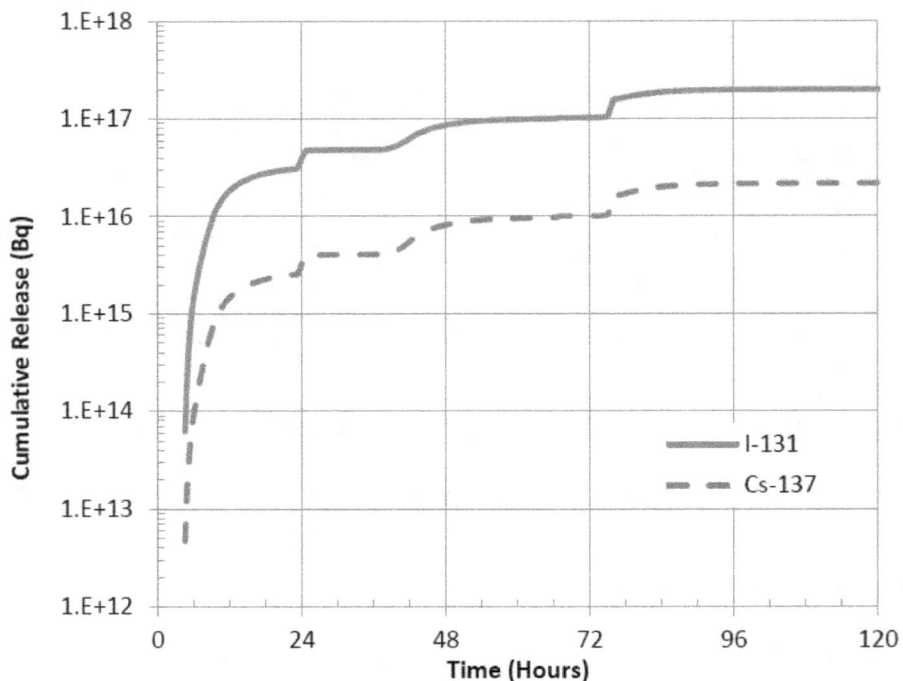

Figure 1-6   RASCAL 4 estimate of the cumulative I-131 and Cs-137 releases for the Fukushima accident

This analysis demonstrates that RASCAL 4 has sufficient flexibility to make reasonable estimates of the Fukushima release rates even with the complexity of the accident scenario.

1.8 References

American National Standard Institute (ANSI/ANS). 1999. "Radioactive Source Term for Normal Operation of Light-Water Reactors," ANSI/ANS-18.1-1999, American Nuclear Society, La Grange Park, IL.

Bateman, H. 1910. "Solution of a System of Differential Equations Occurring in the Theory of Radioactive Decay Transformation, *Proceedings of the Cambridge Philosophical Society*, 15:423–427.

Blevins, R.D. 1984. *Applied Fluid Dynamics Handbook*, Krieger Publishing Company, Malabar, FL.

Chang, R., et al. 2012. "State-of-the-Art Reactor Consequence Analysis (SOARCA) Report: Draft Report for Comment," (Figures 1 and 2), NUREG-1935, U.S. Nuclear Regulatory Commission, Washington, DC.

Eckerman, K.F., et al. 2006. "User's Guide to the DCAL System", ORNL/TM-2001/190, Oak Ridge National Laboratory, Oak Ridge.

Eckerman, K.F., and J.C. Ryman. 1993. "External Exposure to Radionuclide in Air, Water, and Soil," Federal Guidance Report No. 12, EPA-402-R-93-081, U.S. Environmental Protection Agency, Washington, DC.

Inoue, M. 2012. "Overview of Environmental Contamination by Radioactivity Discharged from Fukushima Daiichi NPP," Presented at the Japan/U.S. Department of Energy Workshop on Remediation, February 13–15, 2012.

Institute of Nuclear Power Operations (INPO). 2011. "Special Report on the Nuclear Accident at the Fukushima Daiichi Nuclear Power Station," (Table 7.3), INPO-11-005, Atlanta, GA.

Matsunaga, T. 2012. "Introduction of Fukushima Daiichi Nuclear Power Station Accident," Presented at the International Workshop on Source Term Estimation Methods for the Fukushima Incident, National Center for Atmospheric Research, Boulder, CO, February 22, 2012.

McKenna, T.J., et al. 1996. "Response Technical Manual (RTM)-96," Volume 1, Revision 4, NUREG/BR-0150, U.S. Nuclear Regulatory Commission, Washington, DC.

McKenna, T.J., and J. Giitter. 1988. "Source Term Estimation during Incident Response to Severe Nuclear Power Plant Accidents," NUREG-1228, U.S. Nuclear Regulatory Commission, Washington, DC.

Oak Ridge National Laboratory (ORNL). 1989. "ORIGEN 2: Isotope Generation and Depletion Code, CCC-371," Oak Ridge National Laboratory, Oak Ridge, TN.

Sjoreen A.L., T.J. McKenna, and J. Julius. 1987. "Source Term Estimation Using MENU-TACT," NUREG/CR-4722, U.S. Nuclear Regulatory Commission, Washington, DC.

Soffer, L., et al. 1995. "Accident Source Terms for Light-Water Nuclear Power Plants, Final Report," NUREG-1465, U.S. Nuclear Regulatory Commission, Washington, DC.

U.S. Nuclear Regulatory Commission (NRC). 1990. "Severe Accident Risks: An Assessment for Five U.S. Nuclear Power Plants," NUREG-1150, U.S. Nuclear Regulatory Commission, Washington, DC.

U.S. Nuclear Regulatory Commission (NRC). 1975. "Reactor Safety Study: An Assessment of Accident Risks in U.S. Commercial Nuclear Power Plants," NUREG-75/014 (WASH-1400), U.S. Nuclear Regulatory Commission, Washington, DC.

2. SPENT FUEL SOURCE TERM CALCULATIONS

RASCAL 4 can calculate source terms for three types of spent fuel storage accidents: (1) releases from spent fuel stored in a pool when the water drains from the pool, causing the fuel to become uncovered, overheating the fuel, and causing cladding damage, (2) releases from spent fuel stored in a pool when the fuel is damaged while it is underwater, and (3) releases from spent fuel in a dry storage cask when an accident causes both damage to the cladding of the fuel and loss of the integrity of the cask. The methods used in RASCAL 4 for estimating the release of radioactive materials from damaged spent fuel are based on information in NUREG/CR-6451, "A Safety and Regulatory Assessment of Generic BWR and PWR Permanently Shutdown Nuclear Power Plants" (Travis et al., 1997).

2.1 Basic Method To Calculate Spent Fuel Source Terms

The method to calculate source terms for spent fuel accidents is similar to the method for the nuclear power plant accident source terms. To perform the calculation, RASCAL 4 first calculates the activity of each radionuclide i that is present in the spent fuel (the "inventory I_i"). Second, it calculates the fraction of the inventory of each radionuclide i that is available for release from the spent fuel for the accident being evaluated, the available fraction AF_i. Third, the product of those two terms is multiplied by a reduction factor RF_i (e.g., for reduction by filters). Reduction factors can include several factors working simultaneously. Lastly, RASCAL 4 calculates the source term by radionuclide $S_i(k)$ released to the atmosphere during time step k by multiplying by the leakage fraction $LF(k)$ for time step k. Equation 2-1 describes these calculations as follows:

$$S_i(k) = I_i \times AF_i \times RF_i \times LF(k) \qquad (2\text{-}1)$$

where:

$S_i(k)$ = activity of radionuclide i released to the environment during time step k
I_i = inventory of radionuclide i
AF_i = fraction for radionuclide i available for release
RF_i = reduction factor for radionuclide i
$LF(k)$ = leakage fraction to the environment during the time step

2.2 Spent Fuel Radionuclide Inventories

RASCAL 4 calculates the inventory of each radionuclide in the spent fuel at the time of the discharge from the reactor I_i using the inventories per megawatt thermal in Table 1-1 of this report. The code calculates the core inventory by multiplying by the reactor power. The default reactor power is 100 percent of the rated power, but the user can change it if appropriate.

The inventories are adjusted for burnup. Equation 1-1 in this report adjusts radionuclides with a half-life of longer than 1 year for burnup. The default burnup for spent fuel is 50,000 megawatt day per metric ton of uranium, but the user can adjust the burnup if desired.

RASCAL 4 then calculates the radionuclide inventories in a single fuel assembly by dividing the burnup-corrected core inventory by the number of assemblies in the core (from the reactor database). These inventories are present in a fuel assembly at the time of reactor shutdown.

If the user specifies the spent fuel involved in the accident in terms of "batches," RASCAL 4 calculates the batch inventories by dividing the core inventories by 3. (The code assumes that the batch is one-third of a core.)

The user must define how long ago the fuel was removed from the reactor. RASCAL 4 then calculates the inventories at the time of the accident by correcting the discharge inventory for radiological decay and ingrowth since the last irradiation. The code uses the methods described in Section 1.7 to calculate the decay and ingrowth.

2.3 Fractions of Inventory Available for Release in Spent Fuel Accidents

Table 2-1 shows the fractions of the radionuclide inventories that are available for release during an accident (AF_i).

Table 2-1   Fuel Release Fractions Used in Spent Fuel Accidents*

NUCLIDE GROUP	RELEASE FRACTIONS BY RELEASE TYPE		
	COLD GAP (immediate)	HOT GAP	CLAD BURNING (over 24 hours)
Noble gases (Xe, Kr)	0.4	0.4	1.0
Halogens (I, Br)	3E-3	3E-2	0.7
Alkali metal (Cs, Rb)	3E-3	3E-2	0.3
Tellurium group (Te, Sb, Se)	1E-4	1E-3	6E-3
Barium, strontium (Ba, Sr)	6E-7	6E-6	6E-4
Noble metals (Ru, Rh, Pd, Mo, Tc, Co)	6E-7	6E-6	6E-6
Cerium group (Ce, Pu, Np) and Lanthanides (La, Zr, Nd, Eu, Nb, Pm, Pr, Sm, Y, Cm, Am)	6E-7	6E-6	2E-6

*Source: The information in Table 2-1 is from Table 3.2 in NUREG/CR-6451, rounded to one significant figure. The zirconium clad burning release fractions are the geometric mean of the high and low fractions.

2.3.1 Spent Fuel Pool Water Drained

Spent fuel in a spent fuel pool must remain covered with water or otherwise cooled to remove decay heat, or the zirconium cladding may heat up and undergo rapid oxidation or "burning" that will propagate and eventually spread to all assemblies in the pool. Thus, the inventory I_i used to calculate the release is the entire inventory in the pool.

The user specifies the number of batches that are in the pool (the default is 10). In addition, the user specifies the date on which the most recent batch of spent fuel was last irradiated. By assuming a refueling interval of 1.5 years, RASCAL 4 can set the ages of each batch and apply a radiological decay correction to each one separately.

The calculated radionuclide release is not sensitive to the refueling time interval. Iodines and noble gases have essentially disappeared from all batches except possibly the last batch out of the reactor. For the older batches, the cesium-137 (Cs-137) is by far the most significant radionuclide, and it decays very slowly with time because its half-life is more than 30 years. No provision exists for changing the refueling interval or handling fuel from more than one reactor.

If the cladding burns, radioactive materials may be released from the fuel using the release fractions in Table 2-1. The RASCAL 4 model assumes that clad burning will not start earlier than 2 hours after the fuel is uncovered and not otherwise cooled, such as from sprays or steam cooling.

The user enters the time that the fuel is uncovered and not otherwise cooled. Next, if applicable, the user enters the time when the fuel is recovered or cooled by sprays or steam cooling. The user also enters the time at which radionuclide release from the fuel (clad burning) starts. This time cannot be earlier than 2 hours after the fuel is uncovered. RASCAL 4 assumes that it takes 24 hours for the releases specified in Table 2-1 for clad burning to occur. Thus, one ninety-sixth of the total release fraction listed in Table 2-1 will occur during each 15-minute time step. The release from the fuel terminates when the fuel is recovered, when it is otherwise cooled by sprays or steam cooling, or when the 24 hours have elapsed.

2.3.2 Fuel Damaged Underwater

Determination of the inventory I_i is made from the user specification of the number of fuel assemblies that were damaged. The user also specifies the last date of irradiation for the damaged assembly that was most recently placed in the pool. RASCAL 4 assumes that all damaged fuel has been stored for the same length of time, and the code applies a radiological decay correction to the inventory.

RASCAL 4 assumes that fuel that is mechanically damaged underwater remains cold but that it experiences cladding failure. The available fractions AF_i are the cold gap release fractions in Table 2-1.

2.3.3 Release from a Dry Storage Cask

The code may assume that release from a dry storage cask accident occurs if an accident causes damage to the cladding of the fuel stored in the cask and if it also causes loss of the integrity of the fuel cask.

The user defines the number of fuel assemblies involved in the accident by selecting the type of fuel cask or by entering the number of assemblies directly. If the user specifies the cask type, the calculations can only be done for a single cask. The fuel damage option allows the user to specify what percentage of the stored fuel has sustained the damage. Determination of the inventory I_i is made from the user specification of the numbers of fuel assemblies that were damaged. The user also specifies the last date of irradiation for the damaged assemblies. RASCAL 4 assumes that all damaged fuel in the cask has been stored for the same length of time, and the code applies a radiological decay correction to the inventory.

For mechanical damage to cladding and the cask, the assemblies would be damaged without heating. Therefore, the available fractions AF_i will be the cold gap release fractions in Table 2-1. If cooling is lost for longer than the thermal limit, RASCAL 4 uses the hot gap release fractions from Table 2-1. The code does not assume the occurrence of any release if cooling is lost for less than 24 hours or if the cask is engulfed in a fire because the casks are designed to maintain their integrity in those situations.

No reduction mechanisms (e.g., sprays or filters) modeled would reduce the amount of activity released to the environment. The user specifies the duration of the release and the percentage of the release rate.

2.4 Release Pathways and Reduction Factors

For all spent fuel accidents, the RASCAL 4 user specifies the times for the start and the end of the release. RASCAL 4 calculates a decay correction for each batch of fuel using its decay methodology described in Appendix A and then applies the decay correction to account for the time from shutdown to the start of release. In addition, the code includes decay and ingrowth during the release period in determining the release rates for spent fuel pool accidents. However, unless the spent fuel pool contains fresh fuel, changes in the release rate during the release because of decay and ingrowth are negligible. Because spent fuel in dry cask storage has been out of the reactor for a sufficient amount of time, decay and ingrowth during a release would be negligible. Therefore, RASCAL 4 does not include decay and ingrowth during the release in determining release rates.

For spent fuel pool accidents in which the fuel is uncovered, RASCAL 4 assumes that the release from the spent fuel passes into a building. Fission product reduction caused by scrubbing by the water in the pool does not occur because the pool water is not covering the fuel. The code assumes that the buildings are not airtight; therefore, the holdup time in the building will be negligible compared to the long release period from the fuel. If the release pathway to the environment passes through filters, the code applies a reduction factor RF_i of 0.01 to all radionuclides except noble gases.

For spent fuel damaged underwater in a spent fuel pool, RASCAL 4 applies a reduction factor RF_i of 0.01 to all radionuclides except noble gases to account for scrubbing by the water in the pool. This factor is in addition to reduction by building filters.

Dry cask storage accidents do not require any reduction mechanisms to reduce the amount of activity released.

2.5 Leakage Fractions

For spent fuel accidents involving uncovered fuel, RASCAL 4 ignores building holdup because it is negligible compared to the release period from the fuel. The leakage fraction LF to the environment is therefore one twenty-fourth of the fraction in Table 2-1.

For spent fuel accidents involving fuel damaged underwater, the leakage fraction LF from the building is specified in terms of percentage per hour with a maximum rate of 100 percent per hour.

For fuel casks stored outdoors, the leakage fraction LF from the cask is specified in terms of percentage per hour with a maximum rate of 100 percent per hour. A user would normally select 100 percent per hour to indicate a very fast transfer rate to the environment.

2.6 Reference

Travis, R.J., R.E. Davis, E.J. Grove, and M.A. Azarm. 1997. "A Safety and Regulatory Assessment of Generic BWR and PWR Permanently Shutdown Nuclear Power Plants," NUREG/CR-6451, BNL-52498, U.S. Nuclear Regulatory Commission, Washington, DC.

3. FUEL CYCLE AND MATERIALS SOURCE TERM CALCULATIONS

RASCAL 4 can estimate the amount of radioactive (or hazardous) material released based on a wide variety of potential radiological accident scenarios. The source term calculations performed that pertain to fuel cycle facility and material accidents can be generally categorized as (1) fuel cycle facility and uranium hexafluoride (UF_6) accidents, (2) uranium fires and explosions, (3) criticality accidents, and (4) isotopic releases (e.g., transportation and materials).

3.1 Basic Method To Calculate the Source Term

The method used to calculate source terms for fuel cycle and material accidents is similar to the method for the nuclear power plant accident source terms. RASCAL 4 calculates source terms by time steps. For each time step, RASCAL 4 first calculates the activity that is present (the "inventory I_i"). Second, RASCAL 4 calculates the fraction of the inventory that is available for release for the accident described, the available fraction AF_i. Third, the product of those two terms is multiplied by a reduction factor RF (e.g., for reduction by filters or containment sprays). Reduction factors can include several factors working simultaneously. Last, RASCAL 4 calculates the source term $S(k)$ released to the atmosphere during time step k by multiplying by the leakage fraction $LF(k)$ for time step k. Equation 3-1 describes these calculations as:

$$S_i(k) = I_i(k) \times AF_i(k) \times RF_i(k) \times LF(k),$$ (3-1)

where:

> $S_i(k)$ = activity of radionuclide i released to the environment during time step k
> $I_i(k)$ = inventory of radionuclide i at time step k
> $AF_i(k)$ = available fraction for release for radionuclide i during time step k
> $RF_i(k)$ = reduction factor for radionuclide i during time step k
> $LF(k)$ = leakage fraction to the environment during time step k

The time steps may be of varying length. A source term time step starts whenever the user changes any of the time-dependent data or every 15 minutes, whichever is less. Time steps may be no less than 1 minute and must be an integral number of minutes. Before passing the source term to the atmospheric transport model, RASCAL 4 converts the source term time steps into 15-minute time steps that start on the hour because the atmospheric transport model requires that regularity. With one exception, RASCAL 4 does not calculate radioactive decay and daughter ingrowth for fuel cycle facility and material accidents. Criticality accidents are the exception. RASCAL 4 calculates decay and ingrowth both before and during criticality accident releases. These calculations are made using the methodology described in Appendix A.

3.2 UF$_6$ Releases from Cylinders

3.2.1 Starting Inventory

The user can describe the inventory I of UF_6 available for release from cylinders in the following two ways:

(1)   The user specifies the number and type of cylinders releasing their contents. A simulation allows only one type of cylinder. Table 3-1 lists the cylinder types available and the mass of UF_6 each one contains. RASCAL 4 assumes that each cylinder is filled to its capacity. The total starting

inventory is the sum of the number of each type of cylinder times the amount of UF$_6$ in that type of cylinder.

(2)   The user specifies a total mass of UF$_6$ in cylinders that can be released.

Table 3-1   UF$_6$ Cylinder Inventories

CYLINDER TYPE	AVAILABLE INVENTORY OF UF$_6$ (kg)
2.5 ton (30A, 30B)	2,277
10 ton (48A, 48X)	9,539
14 ton (48Y, 48G, 48F, 48H)	12,338

3.2.2 Release Fractions and Release Rates

The user next selects the UF$_6$ form (liquid, solid, or solid in a fire) and the type of cylinder damage that leads to the release (rupture or valve/pigtail failure). Specifying these serves only to set the default release fractions and release rates shown in Table 3-2.

Table 3-2   Default Release Fractions and Rates Based on UF$_6$ Form
and Cylinder Damage Type*

	CYLINDER RUPTURE		VALVE OR PIGTAIL FAILURE	
FORM OF UF$_6$	RELEASE FRACTION	RELEASE RATE (kg/s)	RELEASE FRACTION	RELEASE RATE (kg/s)
Liquid	0.65	32	See Table 3-3	4
Solid	1.00	0	1.00	0
Solid in fire	1.00	8	1.00	1

*Source: U.S. Nuclear Regulatory Commission (NRC) reports entitled, "RTM-96 Supplement for the Paducah Gaseous Diffusion Plant," issued 1997 (NRC, 1997a), and "RTM-96 Supplement for the Portsmouth Gaseous Diffusion Plant," issued 1997 (NRC, 1997b).

Earlier versions of RASCAL (Version 3.0.5 or previous) used the release rate for the total UF$_6$ inventory, not per cylinder. Thus, if the user wanted to use 3 cylinders and to have each cylinder leaking at 10 kilograms per second (kg/s), he or she would have had to enter a leak rate of 30 kg/s. In RASCAL 4, the release rate is per cylinder. However, the user may enter only one release rate, and the code will use it for each cylinder.

In one special case, the UF$_6$ is liquid, and a valve or pigtail failure causes the release. In this case, RASCAL 4 uses the valve location to set the maximum release fraction. Table 3-3 shows the relationship between the valve position and release fraction. (The code assumes that all cylinders have the same maximum release fraction and release rate.)

Table 3-3 UF$_6$ Release Fractions Based on Valve Location*

VALVE POSITION	MAXIMUM RELEASE FRACTION
360°—top	0.3870
270°—side	0.5528
180°—bottom	0.9222

*Source: RASCAL 4 computed the release fractions using data taken from
Table 22 in NUREG/CR-4360, "Calculational Methods for Analysis of
Postulated UF$_6$ Releases," Volume 1, issued 1995 (Williams 1995).

The mass of UF$_6$ available for release is the starting inventory *I* times the available fraction *AF*.

3.2.3 Release Pathways

RASCAL 4 has three possible release pathways for UF$_6$ cylinder releases: (1) direct to the atmosphere, (2) through a building, or (3) through filters. For some pathways, the code assumes that the UF$_6$ will be fully converted to hydrogen fluoride (HF) and uranyl fluoride (UO$_2$F$_2$) before entering the atmosphere. Table 3-4 lists the situations in which UF$_6$ is converted before release.

Table 3-4 Pathways with and without UF$_6$ Conversion before Release to the Atmosphere

PATHWAY	NO CONVERSION OF UF$_6$	COMPLETE CONVERSION TO HF AND UO$_2$F$_2$
Direct		
- Liquid	X	
- Solid	X	
- Solid in a fire	X	
Through a building		X
Through filters		X

UF$_6$ is converted as 1 kg of UF$_6$ = 0.88 kg UO$_2$F$_2$ + 0.23 kg HF. To determine the uranium activity from the uranium mass and enrichment, RASCAL 4 converts the mass of UF$_6$ to activity using the enrichment level and the specific activity. Section 3.9 describes this conversion.

The direct to atmosphere pathway assumes that all material released from the cylinder(s) enters the atmosphere without being acted upon by any reduction mechanisms. The leak rate to the atmosphere is the leak rate from the cylinder(s). The UF$_6$ available for release is divided by the release rate to determine how many time steps are required for completion of the release. The UF$_6$ is released at a constant rate until it is exhausted.

Releases through a either a building or filters allow the user to specify a release fraction for the HF and the UO$_2$F$_2$, a building air exchange rate (changes per hour), and a start and end time for the release. This release fraction is different from that described earlier. These numbers represent a reduction because of the building or filters; the previous was a reduction by the cylinder.

The computation of radiological decay is not done in UF_6 accident scenarios.

3.3 UF_6 Releases from Cascade Systems

The UF_6 cascade release accident type is available only for the Portsmouth and Paducah gaseous diffusion plants (GDPs). In addition, the cascade release source term option is available only for certain buildings of those facilities. Tables 3-5 and 3-6 list the building names and information about the default inventories and release rates.

Table 3-5 Paducah GDP Buildings and Default Inventory and Release Rates*

BUILDING NAME	CELLS PER UNIT	NUMBER OF UNITS	AVERAGE CELL INVENTORY (lb)	RELEASE RATE (lb/s)
C-331	10	4	4,400	130
C-333	10	6	9,500	130
C-335	10	4	4,600	130
C-337	10	6	8,400	130
C-310	10	1	150	130

*Source: RTM-96 supplement for the Paducah GDP (NRC, 1997a).

Table 3-6 Portsmouth GDP Buildings and Default Inventory and Release Rates*

BUILDING NAME	CELLS PER UNIT	NUMBER OF UNITS	AVERAGE CELL INVENTORY (lb)	RELEASE RATE (lb/s)
X-326	20	2.5	1,000	130
X-330	10	11	5,000	130
X-333	10	8	5,000	130

*Source: RTM-96 supplement for the Portsmouth GDP (NRC,1997b).

3.3.1 Starting Inventory

The user may enter the starting UF_6 inventory I (1) directly as a total mass of UF_6 available for release or (2) as the mass of UF_6 per cell and the number of units or cells in the cascade that are involved in the release. Entering the number of cells gives the amount of material available as the product of the number of cells times their inventory. Entering the number of units gives the amount of material as the cell inventory times the cells per unit times the number of units.

This starting inventory I is multiplied by the user-entered fraction available for release from the cascade to the building. This value has a default of 1.0, which indicates that all material is released into the building. Values less than 1.0 can be used to represent material removed by the structure caused by natural processes. The user also enters the rate at which the material escapes from the cascade into the building. Each building has a default release rate that can be changed by the user.

3.3.2 Release Pathway

The release pathway is based on two building configurations: (1) summer and (2) winter.

In the summer configuration, RASCAL 4 assumes that the building is sufficiently open to the atmosphere (hot inside with all doors and windows open) and that the released UF_6 has essentially an unobstructed path to the outside. This UF_6 is released to the atmosphere at the defined cascade release rate. An application of fractions is not available for release or start and end of release. In addition, a conversion before release to HF and UO_2F_2 cannot be done.

In the winter configuration, the UF_6 enters the building at the defined release rate and is converted as 1 kg of $UF_6 = 0.88$ kg $UO_2F_2 + 0.23$ kg HF.

To determine the uranium activity from the uranium mass and enrichment, RASCAL 4 converts the mass of UF_6 to activity using the enrichment level and the specific activity. Section 3.9 describes this conversion.

Releases through both a building or filters allow the user to specify a release fraction for the HF and the UO_2F_2, a building air exchange rate (changes per hour), and a start and end time for the release. This release fraction is different from that described earlier. These numbers represent a reduction because of the building or filters; the previous was a reduction by the cylinder.

The computation of radiological decay is not done in UF_6 accident scenarios.

3.4 Fires Involving Uranium Oxide

Uranium oxide fires may occur in several different types of facilities. In the milling of uranium ore, a fire can occur in a drum of milled ore or during the process of extracting solvent. After the ore is milled, the production of reactor fuel begins with creating a powder from the uranium dioxide (UO_2). Both wet and dry processes are used to produce this powder. Uranium-oxide-contaminated waste can be stored in several forms, and any of these can be involved in a fire.

3.4.1 Inventory and Fractions Available for Release

The user first selects one of five locations for the fire and specifies additional conditions. This defines the default fraction available for release *AFs* and respirable fractions *RFs*. The *AFs* and *RFs* are considered to be conservative. The *RF* is the fraction of the material released that is expected to be inhaled. The material is defined as all vapors and any particulate material that has a diameter of <10 micrometers (μm). (Note that the source term calculation does not use the *RFs*. The calculation of inhalation dose uses them to reduce the amount of material inhaled.) Table 3-7 shows the default values for the *AFs* and *RFs* (U.S. Department of Energy (DOE)-HDBK-3010-94, "DOE Handbook: Airborne Release Fractions/Rates and Respirable Fractions for Nonreactor Nuclear Facilities: Analysis of Experimental Data," Volume 1(DOE, 1994)).

Table 3-7   Fractions Available for Release and Respirable Fractions Used in Uranium Oxide Fires*

LOCATION OF FIRE	CONDITION	AF	RF
Production process	Dry process	1×10^{-3}	1
	Wet process	3×10^{-5}	1
HEPA filter	At high temperature	1×10^{-4}	1
	Failure	1	1
Incinerator exhaust		4×10^{-1}	1
Waste fire	Solid packaged in drums	5×10^{-4}	1
	Solid loosely packed	5×10^{-2}	1
	Combustible liquid	3×10^{-2}	1
	Noncombustible liquid	2×10^{-3}	1
Uranium mill	Drum in a fire	1×10^{-3}	1
	Solvent extraction	3×10^{-2}	1

*Source: DOE, 1994.

Next, the user specifies the mass of the UO_2 material at risk and specifies a uranium-enrichment level.

RASCAL 4 calculates the uranium mass by first multiplying the mass of UO_2 by 0.88 (the ratio of atomic weights for uranium and UO_2). Then, the code converts the uranium mass to activity based on enrichment as described in Section 3.9. The source term available for release is the product of this activity times the fraction available for release.

3.4.2 Release Pathway

Releases outside the building have no further reductions. The release rate to the atmosphere is constant at a rate set by dividing the available activity for release (curies) by the release duration specified.

Releases inside the building are similar but reduce the activity available for release before calculating a release rate to the atmosphere. The activity is multiplied by a reduction factor of 0.5 for unfiltered releases and multiplied by 0.01 for filtered releases.

3.5 Explosions Involving Uranium Oxide

Uranium oxide explosions are characterized as (1) those caused by the detonation of high explosives in contact with the material, (2) those caused by a fire (deflagration), and (3) those caused by a sudden pressure change in the material container (venting). The UO_2 in the explosion may be in liquid, solid, or powder form, or it simply may be surface contamination. Table 3-8 lists the default release fractions and respirable fractions for the different explosion types and material forms.

Table 3-8 Fractions Available for Release and Respirable Fractions Used in Uranium Oxide Explosions*

EXPLOSION CHARACTERISTICS	MATERIAL FORM OF THE URANIUM OXIDE	FRACTION AVAILABLE FOR RELEASE	RESPIRABLE FRACTION
Detonation	Liquid	1	1
	Solid	1	2×10^{-1}
	Powder	1	2×10^{-1}
	Surface contamination	1×10^{-3}	1
Deflagration	Liquid	1×10^{-6}	1
	Solid	0	0
	Powder	5×10^{-3}	3×10^{-1}
	Surface contamination	1×10^{-3}	1
Venting	Liquid	2×10^{-3}	1
	Solid	0	0
	Powder	1×10^{-1}	7×10^{-1}
	Surface contamination	1×10^{-3}	1

*Source: DOE, 1994.

First, the user selects an explosion type and a material form. This defines the default fraction available to be released and the respirable fraction. Next, the user specifies the mass of the UO_2 material at risk and specifies a uranium-enrichment level.

RASCAL 4 calculates the uranium mass by first multiplying the mass of UO_2 by 0.88 (the ratio of atomic weights for uranium and UO_2). Then the code converts the uranium mass to activity based on enrichment as described in Section 3.9. The source term available for release is the product of this activity times the release fraction.

3.5.1 Release Pathway

Releases outside the building have no further reductions. The release rate to the atmosphere is constant at a rate set by dividing the available activity for release (curies) by the release duration specified.

Releases inside the building are similar but reduce the available activity to release before calculating a release rate to the atmosphere. The activity is multiplied by 0.5 for unfiltered releases and multiplied by 0.01 for filtered releases.

3.6 Criticality Accidents

A criticality accident results from the uncontrolled release of energy from an assemblage of fissile material. In RASCAL 4, the user can model a criticality accident using the physical system scenarios in NUREG/CR-6410, "Nuclear Fuel Cycle Facility Accident Analysis Handbook," (Science Applications International Corporation (SAIC), 1998) or using the criticality data that he or she has entered directly.

Table 3-9 lists the physical systems modeled and the assumed number of fissions in the first burst and the total yield. The user selects whether to model a single burst or multiple bursts. The assumption is that the bursts come at 10-minute intervals and continue for 8 hours (a total of 48 bursts).

Table 3-9 Fission Yields Used in Criticality Calculations*

SYSTEM MODELED IN THE SCENARIO	INITIAL BURST YIELD (fissions)	TOTAL YIELD (fissions)
Solution <100 gallons	1×10^{17}	3×10^{18}
Solution >100 gallons	1×10^{18}	3×10^{19}
Liquid/powder	3×10^{20}	3×10^{20}
Liquid/metal pieces	3×10^{18}	1×10^{19}
Solid uranium	3×10^{19}	3×10^{19}
Solid plutonium	1×10^{18}	1×10^{18}
Large storage arrays below prompt critical	None	1×10^{19}
Large storage arrays above prompt critical	3×10^{22}	3×10^{22}

*Source: SAIC, 1998.

The physical systems method uses Equation 3-2 to calculate the number of fissions in a burst F_B for all except the first burst as:

$$F_B = \frac{F_t - F_I}{(48 - 1)},$$

(3-2)

where:

F_T = the total yield (fissions) of the criticality (Column 3 in Table 3-9)

F_I = the yield (fissions) of the initial burst (Column 2 in Table 3-9)

The user-defined method requires setting the following parameters:

- number of fissions (F_I) in the first burst
- number of fissions (F_B) in the subsequent bursts
- burst interval in minutes
- duration of the criticality

This user-defined method assumes that a multiple-burst event will end after 48 bursts, irrespective of the burst interval.

For both methods of defining the fission yield of the criticality, the user defines fractions available for release for noble gases, iodines, and other nuclides that have default values of 1.0, 0.25, and 0.0005, respectively.

The user also defines shielding thicknesses for steel, concrete, and water. Those thicknesses are used to calculate the reduction in the neutron and gamma prompt shine dose because of shielding.

Table 3-10 lists the assumed amounts of each radionuclide released per 1×10^{19} fissions (SAIC, 1998). These values are based on ORIGEN 2 calculations (ORNL, 1989).

To calculate the source term, RASCAL 4 first determines the initial activity of each radionuclide present as the product of the yield of the initial burst (F_I) (in 1×10^{19} fissions) and the activity per 1×10^{19} fissions listed in Table 3-10. For each following time step, RASCAL 4 (1) determines whether the criticality is still occurring and whether enough time has passed for the occurrence of one or more subsequent bursts and, if so, adds the appropriate activity as the product of the yield from these burst (F_B) and the activity per 1×10^{19} fissions, (2) reduces the amount of activity for the amount released, and (3) applies the release fractions and radiological decay to the result. A criticality will end when either the total number of allowed bursts have been accounted for or when the "end of criticality" time entered by the user has been reached. If the user selects a release duration that is not long enough to include all 48 bursts, the total activity released will be less than the amount listed in Table 3-10.

Table 3-10 Activity (Curies) Released in Criticality of 1x10^{19} Fissions*

RADIONUCLIDE	ACTIVITY (Ci)	RADIONUCLIDE	ACTIVITY (Ci)
Kr-83m	1.5E+02	I-131	7.3E+00
Kr-85m	8.9E+01	I-132	1.0E+03
Kr-85	1.3E-05	I-133	1.7E+02
Kr-87	1.1E+03	I-134	4.2E+03
Kr-88	6.6E+02	I-135	5.0E+02
Kr-89	4.6E+04	Sr-91	3.2E+02
Xe-133m	1.9E-02	Sr-92	1.2E+03
Xe-133	2.7E-03	Ru-106	2.0E-02
Xe-135m	3.3E+02	Cs-137	1.0E-02
Xe-135	5.2E+00	Ba-139	2.4E+03
Xe-137	2.4E+04	Ba-140	1.1E+01
Xe-138	1.0E+04	Ce-143	1.0E+02

*Source: SAIC, 1998.

3.6.1 Release Pathway

RASCAL 4 assumes that the criticality takes place inside a building. The user selects a leak rate to the atmosphere from this building from the following four choices:

(1)   100 percent per hour (represents ordinary building ventilation)
(2)   50 percent per hour
(3)   10 percent per hour
(4)   4 percent per hour (equivalent to 100 percent per day)

This release rate method releases a fixed fraction of the material per unit time. After the criticality stops, the release rate to the environment decreases exponentially.

The user defines a start and end of the release to the atmosphere, which describes when the radionuclide material generated by the criticality enters the environment.

The user may define reduction factors for noble gases, iodines, and other radionuclides. These are multiplied times the appropriate nuclide activities to reduce the release.

3.6.2 Prompt Shine Dose Calculation

For criticality accidents, the criticality-shine dose is computed with the source term. The shielding thicknesses entered by the user are used only in this calculation.

The dose in rem D_{crit} at 10 feet is computed as follows (NUREG/CR-6504, "An Updated Nuclear Criticality Slide Rule," Volume 2 (Hopper and Broadhead, 1998)):

$$D_{crit} = D_{gamma} + D_{neutron},$$ (3-3)

$$D_{gamma} = 1 \times 10^{-15} \times F_T \times e^{-(0.386 \times S + 0.147 \times C + 0.092 \times W)}, \text{ and}$$ (3-4)

$$D_{neutron} = 1 \times 10^{-14} \times F_T \times e^{-(0.256 \times S + 0.240 \times C + 0.277 \times W)},$$ (3-5)

where:

F_T = the total number of fissions
S = the thickness of steel shielding in inches
W = the thickness of water shielding in inches
C = the thickness of concrete shielding in inches

Doses D at other distances are computed using the inverse square law as:

$$D(x) = \left(\frac{10\text{ ft}}{x^2}\right) \times D(10\text{ ft}),$$ (3-6)

where x is the distance in feet.

3.7 Sources and Material in a Fire

In a fire release, the user enters the amount of each radionuclide present. No release occurs when the fire is not burning, and no other types of reduction are allowed. A fire may start and stop burning only once. The default values for these fire reduction factors are from NUREG/BR-0150, "Response Technical Manual: RTM-96," Volume 1, Revision 4 (McKenna et al., 1996). The user can select fire release fractions by element or the form of the compound, or he or she can enter them directly. Tables 3-11 and 3-12 show the default fire release fractions used. Note that the total amount of activity released also depends on the release duration entered in the isotopic release pathway form. For example, if the release duration is shorter than the duration of the fire, RASCAL 4 reduces the amount of activity released.

Table 3-11   Fire Release Fractions by Compound Form*

FORM OF COMPOUND	RELEASE FRACTION
Noble gas	1.0
Very mobile form	1.0
Volatile or combustible compound	0.5
Carbon	0.01
Semivolatile compound	0.01
Nonvolatile compound	0.001
Uranium and plutonium metal	0.001
Nonvolatile in a flammable liquid	0.005
Nonvolatile in a nonflammable liquid	0.001
Nonvolatile solid	0.0001

*Source: Table F-2 in NUREG/BR-0150 (McKenna et al., 1996).

If the compound form is not known, the user enters the fire release fractions in Table 3-12.
The fire release fraction is the fraction of the isotope released when the material is involved in a fire; it equals the total activity released (curies) divided by the activity involved in a fire (curies).

Table 3-12 Fire Release Fractions by Element[a]

ELEMENT[b]	RELEASE FRACTION [c]	ELEMENT	RELEASE FRACTION	ELEMENT	RELEASE FRACTION	ELEMENT	RELEASE FRACTION
H (gas)	0.5	Se	0.01	I	0.5	W	0.01
C	0.01	Kr	1.0	Xe	1.0	Ir	0.001
Na	0.01	Rb	0.01	Cs	0.01	Au	0.01
P	0.5	Sr	0.01	Ba	0.01	Hg	0.01
S	0.5	Y	0.01	La	0.01	Tl	0.01
Cl	0.5	Zr	0.01	Ce	0.01	Pb	0.01
K	0.01	Nb	0.01	Pr	0.01	Bi	0.01
Ca	0.01	Mo	0.01	Pm	0.01	Po	0.01
Sc	0.01	Tc	0.01	Sm	0.01	Ra	0.001
Ti	0.01	Ru	0.1	Eu	0.01	Ac	0.001
V	0.01	Rh	0.01	Gd	0.01	Th	0.001
Cr	0.01	Ag	0.01	Tb	0.01	Pa	0.001
Mn	0.01	Cd	0.01	Ho	0.01	U	0.001
Fe	0.01	In	0.01	Tm	0.01	Np	0.001
Co	0.001	Sn	0.01	Yb	0.01	Pu	0.001
Zn	0.01	Sb	0.01	Hf	0.01	Am	0.001
Ge	0.01	Te	0.01	Ta	0.001		

[a] Table F-3 in NUREG/BR-0150 (McKenna et al., 1996). The release fraction for ruthenium was changed from the value of 0.01 in NUREG-1140, "A Regulatory Analysis on Emergency Preparedness for Fuel Cycle and Other Radioactive Material Licensees (Final Report)," (McGuire, 1988), to a value of 0.1. NUREG-1140 assumes that ruthenium is nonvolatile (McGuire, 1988). However, research in NUREG/CR-6218, "A Review of the Technical Issue of Air Ingression during Severe Reactor Accidents," (Powers et al., 1994), indicates (in Table 5) that ruthenium starts to become volatile at high temperatures. The ruthenium release fraction of 0.1 is less than the value of 0 5 used in NUREG-1140 for compounds because ruthenium is less volatile than other volatile compounds; it becomes highly volatile only at temperatures not normally reached in building fires. The carbon release fraction is appropriate for carbon compounds other than CO_2. Those compounds deliver most of the dose. The dose conversion factors used for carbon are for those carbon compounds.

[b] If the specific physical form of the nuclide is known, Table 3-11 may be used.

[c] The fire release fraction is the fraction of the isotope released when the material is involved in a fire, and it equals the total activity released (curies); users should divide by the activity involved in a fire (curies).

Fire release fractions are element specific. Tables 3-11 and 3-12 show the fire reduction factors; these factor are from NUREG-1140 (McGuire, 1988)

For all types of isotopic releases, if the user selects release units in mass instead of activity, RASCAL 4 converts the source term to curies using the specific activity of each radionuclide. The user may specify the enrichment level for enriched uranium. The enrichment level for natural uranium is assumed to be 0.7 percent (Table E-5 in NUREG/BR-0150 (McKenna et al., 1996)). Specific activity is computed as described in Section 3.9. For natural and enriched uranium, radiological decay and dose are calculated assuming the properties of uranium-238 (U^{238}) and U^{234}, respectively. U^{234} is used rather than U^{235} because U^{234} has a specific activity approximately 3 orders of magnitude higher than that of U^{235}.

3.8 Isotopic Release Rates and Concentrations

Chapter 1 discusses in detail isotopic release rates and concentration source term types. They are available for all the event types of RASCAL 4 except spent fuel.

3.9 Computing Uranium-Specific Activity from Enrichment

RASCAL 4 calculates the specific activity of uranium using the user-entered value for the enrichment. The code generates a cubic spline using the data points given in Table 3-13. It then evaluates this spline for the given enrichment to provide the specific activity.

Table 3-13   Uranium-Specific Activity for Different Enrichments

ENRICHMENT (percent U^{235} by weight)	SPECIFIC ACTIVITY (μCi/g)
0.0 (depleted)	0.4
4.0	2.4
93.0	110.0

*Source: Table E-5 in NUREG/BR-0105 (McKenna et al., 1996).

3.10 References

Hopper, C.M., and B.L. Broadhead. 1998. "An Updated Nuclear Criticality Slide Rule," Volume 2, NUREG/CR-6504, ORNL/TM-13322, U.S. Nuclear Regulatory Commission, Washington, DC.

McGuire, S.A. 1988. "A Regulatory Analysis on Emergency Preparedness for Fuel Cycle and Other Radioactive Material Licensees (Final Report)," NUREG-1140, U.S. Nuclear Regulatory Commission, Washington, DC.

McKenna, T.J., et al. 1996. "Response Technical Manual: RTM-96," Volume 1, Revision 4, NUREG/BR-0150, U.S. Nuclear Regulatory Commission, Washington, DC.

McKenna, T.J., and J. Giitter. 1988. "Source Term Estimation during Incident Response to Severe Nuclear Power Plant Accidents," NUREG-1228, U.S. Nuclear Regulatory Commission, Washington, DC.

Oak Ridge National Laboratory (ORNL). 1989. "ORIGEN 2: Isotope Generation and Depletion Code, CCC-371," Oak Ridge National Laboratory, Oak Ridge, TN.

Powers, D.A., L.N. Kmetyk, and R.C. Schmidt. 1994. "A Review of the Technical Issue of Air Ingression during Severe Reactor Accidents," NUREG/CR-6218, SAND-94-0731, U.S. Nuclear Regulatory Commission, Washington, DC.

Science Applications International Corporation (SAIC). 1998. "Nuclear Fuel Cycle Facility Accident Analysis Handbook," NUREG/CR-6410, Science Applications International Corporation, Reston, VA.

U.S. Department of Energy (DOE). 1994. "DCE Handbook: Airborne Release Fractions/Rates and Respirable Fractions for Nonreactor Nuclear Facilities: Analysis of Experimental Data," Volume 1, DOE-HDBK-3010-94, U.S. Department of Energy, Washington, DC.

U.S. Nuclear Regulatory Commission (NRC) (1997a), "RTM-96 Supplement for the Paducah Gaseous Diffusion Plant," Pacific Northwest National Laboratory, Richland, WA.

U.S. Nuclear Regulatory Commission (NRC) (1997b), "RTM-96 Supplement for the Portsmouth Gaseous Diffusion Plant," Pacific Northwest National Laboratory, Richland, WA.

Williams, W.R. 1995. "Calculational Methods for Analysis of Postulated UF_6 Releases," Volume 1, NUREG/CR-4360. U.S. Nuclear Regulatory Commission, Washington, DC.

4. TRANSPORT, DIFFUSION AND DOSE CALCULATIONS

RASCAL 4 uses Gaussian models to describe the atmospheric dispersion of radioactive and chemical effluents from nuclear facilities. The U.S. Nuclear Regulatory Commission (NRC) staff frequently use these models in licensing and emergency response calculations because they quickly provide reasonable estimates of atmospheric concentrations, deposition, and doses given relatively limited information on topography and meteorology (e.g., PAVAN (Bander, 1982), XOQDOQ (Sagendorf et al., 1982), and MESORAD (Scherpelz et al., 1986; Ramsdell et al., 1988)). A straight-line Gaussian plume model, TADPLUME, is used near the release point where travel times are short. A Lagrangian-trajectory Gaussian puff model, TADPUFF, is used at longer distances for which temporal or spatial variations in meteorological conditions may be significant.

Although the basic Gaussian models in this version of RASCAL are the same as those used in previous versions, RASCAL 4 has improved algorithms for calculating dispersion parameters, dry deposition velocities, and wet deposition. It also includes an updated treatment of the deposition of iodines. These new algorithms were developed as part of the Hanford Environmental Dose Reconstruction (HEDR) Project (Shipler et al., 1996) and have undergone extensive review.

Chapter 4 describes TADPLUME and TADPUFF. It begins with a short theoretical derivation of Gaussian plume and puff models and then describes the implementation of those models in TADPLUME and TADPUFF. The chapter also describes the dose calculations that are embedded in TADPLUME and TADPUFF. RASCAL 4 expands the dose calculations to include intermediate-phase dose projections for the first and second years and 50 years following a release. The chapter ends with a discussion of the verification and validation TADPLUME and TADPUFF.

Chapter 5 describes UF6PLUME, a straight-line Gaussian plume model developed to handle the complexities associated with releases of uranium hexafluoride (UF$_6$). UF6PLUME includes many of the components of TADPLUME.

4.1 Theoretical Bases for Gaussian Models

Many texts discuss the derivation of the Gaussian models used to describe atmospheric dispersion. Various texts (e.g., Slade,1968; Csanady, 1973; Randerson,1984; and Seinfeld, 1986) provide the bases for the following discussion. They may be consulted for additional detail as desired.

A differential equation called the diffusion equation, in part, governs atmospheric dispersion. With a set of assumptions that can reasonably be applied to atmospheric processes, the diffusion equation has a specific, closed-form algebraic solution that is Gaussian. In one dimension, the solution is

$$\chi(x)/Q = \frac{1}{(2\pi)^{1/2}\sigma} exp\left[-\frac{1}{2}\left(\frac{x-x_o}{\sigma}\right)^2\right], \tag{4-1}$$

where:

$\chi(x)$ = concentration at a distance x-x_o from the center of the concentration distribution x_o (1/m)
Q = amount of material released (Ci or g),
σ = dispersion parameter (m).

Atmospheric dispersion parameters are functions of either distance from the release point or time since release. They may also be functions of atmospheric stability and surface roughness. Numerous atmospheric dispersion experiments provide the data used to evaluate dispersion parameters and to develop methods to predict dispersion-parameter values from readily available data. Draxler (1984) describes a number of these experiments.

4.1.1 Gaussian Puff Model

Using the principle of superposition, the one-dimensional solution of the diffusion equation can be expanded to three dimensions to obtain the basic Gaussian puff model. In a Cartesian coordinate system with x and y axes in a horizontal plane and with z in the vertical plane, the normalized concentration in the vicinity of the puff is:

$$\frac{\chi(x,y,z)}{Q} = \frac{1}{(2\pi)^{3/2} \sigma_x \sigma_y \sigma_z} \, exp\left[-\frac{1}{2}\left(\frac{x-x_0}{\sigma_x}\right)^2\right] exp\left[-\frac{1}{2}\left(\frac{y-y_0}{\sigma_y}\right)^2\right] exp\left[-\frac{1}{2}\left(\frac{z-z_0}{\sigma_z}\right)^2\right], \qquad (4\text{-}2)$$

where χ is the concentration (Ci/m^3 or g/m^3) and Q is the amount of material released (Ci or g).

This equation, when combined with a transport mechanism to move the center of the puff (x_o, y_o, z_o), is a simplified version of TADPUFF, the puff model in RASCAL 4. The dispersion parameters are shown as functions of direction from the puff center. However, most implementations of the puff model, including TADPUFF, assume that the puff is symmetrical in the x and y directions. Hence, x and y may be replaced by the horizontal distance r from the center of the puff.

The form of Equation 4-2 shown is appropriate if the height of the center of the puff is such that vertical dispersion proceeds unimpeded by either the ground or an elevated layer of the atmosphere. Unimpeded vertical dispersion is generally not the case.

Typically, Earth's surface and the top of the atmospheric mixing layer are assumed to be reflective surfaces. When these assumptions are made, the sum of exponential terms that account for reflection replaces the following vertical exponential term

$$exp\left[-\frac{1}{2}\left(\frac{z-z_0}{\sigma_z}\right)^2\right]$$

with

$$\sum_{n=-\infty}^{\infty} \left\{ exp\left[-\frac{1}{2}\left(\frac{2nH-h-z}{\sigma_z}\right)\right]^2 + exp\left[\frac{2nH+h-z}{\sigma_z}\right]^2 \right\},$$

where H is the height of the top of the mixing layer (m), and h is the release height (m). In practice, it is only necessary to include a small number of terms in the summation. RASCAL 4, as in MESORAD (Scherpelz et al., 1986; Ramsdell et al., 1988) and earlier versions of RASCAL, carries out the summation from $n = -2$ to 2. This term can be simplified if one or more of H, h, or z equals zero. For example, if H is large compared to σ_z and if z is zero, the following term may replace the summation:

$$2 \, exp\left[-\frac{1}{2}\left(\frac{h}{\sigma_z}\right)^2\right]$$

At long downwind distances at which the vertical dispersion parameter is the same magnitude as the mixing layer thickness, assuming that material is uniformly distributed in the vertical can simplify the puff model. With this last assumption, the puff model becomes:

$$\chi(r)/Q = \frac{1}{2\pi\sigma_r^2 H}\, exp\left[-\frac{1}{2}\left(\frac{r}{\sigma_r}\right)^2\right],$$ (4-3)

where H is the mixing layer thickness. RASCAL 4 switches to the uniformly mixed model when $\sigma_z > 1.05H$, because the model with reflections and the uniformly mixed model give nearly identical estimates of χ/Q.

4.1.2 Straight-Line Gaussian Plume Models

Puff models represent plumes as a series of puffs. These models calculate concentrations at a point in the plume by adding the concentrations at the point associated with all puffs in the vicinity of the point. In effect, the puff models perform a numerical time integration of concentration as puffs pass by the point. Near a release point, the assumption may be made that the meteorological conditions are constant as the puff moves from the source to the receptor. Integration of the puff model can be done analytically to give a plume model by assuming that the wind speed is much greater than zero and that the point for which the concentration is to be calculated is sufficiently far downwind so that the change in dispersion parameters with distance as puffs pass the point can be neglected.

Assuming that the x axis is aligned with the mean transport direction and that the mean wind speed is u, the exposure (concentration multiplied by time) at a distance x downwind of the release point during passage of a puff is given by:

$$E(x,0,0) = \int_{t=-\infty}^{t=\infty} \frac{QF_yF_z}{(2\pi)^{3/2}\,\sigma_x(x)\sigma_y(x)\sigma_z(x)}\, exp\left[-\frac{1}{2}\left(\frac{x-ut}{\sigma_x(x)}\right)^2\right]dt,$$ (4-4)

where:

> $E(x,0,0)$ = exposure (Ci-s/m^3 or g-s/m^3),
> Q = amount of material in the puff (Ci or g),
> F_y, F_z = lateral and vertical exponential terms as shown above
> x = downwind distance (m) at which χ, σ_x, σ_y, and σ_z are evaluated
> u = wind speed (m/s),
> t = time (s) of puff passage at x (puff center is at x at $t = 0$)

On integration, the model becomes:

$$E(x,y,x)/Q = \frac{F_yF_z}{2\pi u\sigma_y\sigma_z}$$ (4-5)

Equation 4-5 is a simplified version of the equation used in TADPLUME, the straight-line Gaussian model in RASCAL 4. Then, assuming that the material is released over some short time period T instead of as a puff, both E and Q are divided by T to give the average concentration in a plume χ, the release rate Q', respectively, to give:

$$\frac{\chi(x,y,x)}{Q'} = \frac{F_yF_z}{2\pi u\sigma_y\sigma_z}$$ (4-6)

4-3

The straight-line Gaussian plume model for ground-level concentrations from ground-level releases is frequently given as:

$$\chi/Q' = \frac{1}{\pi u \sigma_y \sigma_z} exp\left[-\frac{1}{2}\left(\frac{y}{\sigma_y}\right)^2\right], \hspace{2cm} (4\text{-}7)$$

where F_y in Equations 4-5 and 4-6 is the exponential term in Equation 4-7. When the release and receptor are at ground level and when H is large, the sum of exponential terms that comprise F_z has a value of 2. Hence, the constant 2 in Equations 4-5 and 4-6 does not appear in Equation 4-7.

Another assumption that deserves comment is that the meteorological conditions are assumed to be horizontally homogeneous and stationary. Under this assumption, the wind direction and speed responsible for transporting the plume from the release point to the receptor and the turbulence responsible for diffusion do not change with location throughout the model domain. In addition, the meteorological conditions do not change as a function of time during the release and time required for transport. Together, these assumptions constrain the usefulness of the straight-line plume model to estimating concentrations and doses at receptors near the release point for releases that are a few minutes to about 1 hour in duration. Other models should be used for longer duration releases and for longer distances.

4.1.3 Treatment of Calm Winds

The Gaussian puff model behaves well in calm winds. If the dispersion parameters are a function of time as they are in many models, the material in the puff continues to disperse even though it is not moving. If the dispersion parameters are calculated as the function of travel distance as they were in previous versions of RASCAL, dispersion ceases during calm winds, and the material distribution remains unchanged as long as the wind is calm. In either case, deposition, depletion, exposures, and doses are calculated just as they are during windy conditions.

The straight-line Gaussian plume model in Equation 4-5 tends to overestimate concentrations and doses during low wind speed conditions and becomes undefined for calm wind conditions because wind speed is in the denominator. This behavior occurs because the derivation of the straight-line Gaussian plume model assumes that the wind speed is significantly greater than zero, thus eliminating a portion of the solution of the dispersion equation that deals with low wind speed diffusion. To compensate for the missing part of the solution, straight-line models may assume a wind speed of 0.5 to 1 meter per second (m/s) when calm winds are encountered.

When the wind speed falls below 0.447 m/s (1 mile per hour), TADPLUME switches from the standard Gaussian plume model previously described to the model used in TADPUFF. As long as the wind speed remains below 0.447 m/s, puffs are released at 5-minute intervals. All puffs are retained and continue to grow radially until the wind speed increases above 0.447 m/s. Dispersion, deposition, and exposures are calculated using time-based dispersion coefficients as they are in TADPUFF. When the wind speed increases, the puffs are deleted, and TADPLUME reverts to the standard Gaussian plume model.

While the wind speed is below 0.447 m/s, an inconsistency exists between concentrations and dose fields calculated by TADPLUME and TADPUFF. The air and ground concentrations and the doses calculated by TADPUFF, which are presented on the Cartesian grid described below, will show an elongated plume as the puffs drift downwind. In contrast, the pattern for air and ground concentrations and doses calculated by TADPLUME, which are presented on the polar grid described below, will show a circular pattern. The pattern on the Cartesian grid shows concentration and dose estimates based on the most recently available wind direction. This wind direction may or may not be representative of the plume motion. The pattern on the polar grid shows concentrations and dose estimates as circular, assuming no mean motion of the plume. In reality, the magnitudes of the concentrations and doses in the environment are likely be between those shown on the polar and Cartesian grids with uncertain locations of the maximum values.

4.1.4 Model Domains

TADPLUME and TADPUFF use different model domains. The TADPLUME domain consists of a polar grid with receptor nodes on circles at 10-degree intervals at eight radial distances. Users may adjust distances to the circles to suit the problem at hand. The TADPUFF domain consists of a square Cartesian grid with receptor nodes uniformly spaced throughout the domain. The polar grid has a higher node density near the release point than that of the Cartesian grid; conversely, the Cartesian grid has a higher node density in the far field than the polar grid.

In general, the receptor nodes for the two grids do not coincide. This fact leads to apparent discrepancies in the doses reported in the maximum value tables in the model output for the two models for wind directions other than north, east, south, or west. The doses reported for TADPLUME are for the plume centerline at each distance. The doses reported for TADPUFF are the highest doses calculated at nodes at about a nominal distance (e.g., 5 miles). The node with the highest dose may or may not be on the plume centerline and may be nearer to, or farther from, the release point than the nominal distance. To directly compare doses calculated by the two models, the user should select the wind direction for the period of calculations as either north, east, south, or west. For these wind directions, both TADPLUME and TADPUFF calculate plume centerline concentrations and doses.

4.2 Transport

Atmospheric transport refers to the movement of material from the source to downwind receptors. The model revisions in RASCAL 4 have not changed the transport algorithms from those used in previous versions. The following two sections describe the treatment of atmospheric transport in RASCAL.

4.2.1 TADPLUME Transport

TADPLUME is a straight-line Gaussian model. As this name implies, the model assumes straight-line transport based on the wind direction at the time and place of release. TADPLUME rounds the wind direction to the closest 10 degrees as it calculates the transport direction to ensure that the axis of the plume passes directly over receptors. Straight-line Gaussian models commonly do not generally consider transit time in determining when material arrives at receptors; material arrives at receptors at the time of release. As a result, dose rates calculated by TADPLUME cannot be used to estimate the time of arrival of a plume at a receptor and are not likely to correspond with dose rates measured in the field.

An exception to the general rule occurs when the wind speed at the release point is less than 0.447 m/s. Under this condition, no transport occurs, but dispersion, which is a function of time, does occur. The model determines arrival of material at the receptors by the time required for dispersion to the receptors to take place. Material will arrive at each ring of receptors at a different time.

Although travel time generally is not considered in plume transport, o calculate the decay of radionuclides between the source and the receptors. The model also uses transit time to calculate depletion of material in the plume caused by dry and wet deposition. The model calculates decay at 5-minute intervals and depletion for the full transit time.

4.2.2 TADPUFF Transport

Unlike TADPLUME, TADPUFF explicitly accounts for transit time in all calculations because the model tracks the movement of individual puffs and calculates concentrations and doses based on puff positions. As a result, dose rates calculated by TADPUFF may be used to estimate the time of arrival of a plume and may be compared to dose rates measured in the field. The model calculates decay and ingrowth of radionuclides and depletion of the puffs as a result of dry and wet deposition at 5-minute intervals.

TADPUFF differs significantly from TADPLUME in that neither the wind data nor the wind fields are modified to force the centers of puffs to pass directly over the receptor nodes. As a result, when the wind direction is constant, TADPUFF may not calculate centerline concentrations and doses. However, as time goes by and atmospheric conditions (wind direction, wind speed, stability, mixing layer thickness, and precipitation) change, TADPUFF will give more realistic concentration and dose patterns than TADPLUME will. In addition TADPUFF will give more realistic concentration and dose patterns than TADPLUME will when topography modifies the winds because the wind fields used by TADPUFF may be modified to account for topography. Modification of wind fields to account for topography is done by the meteorological data processor and is described in Section 6.5.2. The wind data used by TADPLUME are not modified to account for topography.

The wind at the center of the puff controls the movement of the puff as it moves through the model domain. TADPUFF represents the spatial variation of winds by two-dimensional fields of vectors that give the direction and speed of puff movement. The meteorological model discussed in Chapter 6 prepares these fields, and they are updated at 15-minute intervals based on the available wind data.

The calculation of puff movement is done using a six-step sequential process:

(1)   Make an initial estimate of the direction and speed of the puff movement given the current puff position and height aboveground using bilinear interpolation (Press et al., 1986) of the vectors at the nearest nodes of the field.

(2)   Make an initial estimate of the puff position at the end of the period using the initial estimates of direction and speed.

(3)   Make a second estimate of the direction and speed of puff movement using the estimated puff position at the end of the period.

(4)   Make a second estimate of the puff position at the end of the period using the estimate of direction and speed from Step 3.

(5)   Average the end points calculated in Steps 2 and 4.

(6)   Calculate the final estimate of direction and speed of puff movement using the puff's initial position and the average end point calculated in Step 5.

The actual puff movement for the period may take place in one or several steps. The model adjusts the step size to ensure adequate accuracy in the integration of concentrations that takes place at receptors. Errors in the integration should be less than 5 percent at typical wind speeds. Larger errors may occur near the release point in high wind speed conditions because the minimum step size is 15 seconds. These larger errors should not be a problem because plume model output should be used for receptors near the release point.

The vector fields prepared by the meteorological program are for a height of 10 meters aboveground. The model uses these vectors for puffs that represent ground-level releases. If the actual release height is greater than 12 meters, the model uses a wind speed profile to adjust the transport speed from 10 meters to the puff transport height. The profile used to adjust the wind speed considers both surface friction and atmospheric stability. (See Sections 6.4–6.6 of Panofsky and Dutton (1984).)

4.3 Dispersion Parameters

Previous versions of RASCAL calculated the horizontal and vertical dispersion parameters (σ_y and σ_z), used in TADPLUME and TADPUFF, using an empirical curve derived from the results of a large number of dispersion experiments conducted in the 1950s and 1960s. The experiments, which were conducted over relatively flat terrain, typically involved tracer releases ranging from about 10 minutes to 1 hour in duration with ground-level concentration measurements at distances ranging from 100 meters to several kilometers. Only a few direct measurements of vertical dispersion parameters (σ_z) were made. Consequently, dispersion models estimated vertical dispersion parameters using measured values of the horizontal dispersion parameter and measured concentrations. Dispersion parameters have been summarized in many forms. Perhaps the best known summary is the set of dispersion parameter curves called the Pasquill-Gifford curves (Gifford, 1976).

RASCAL 4 uses dispersion parameters that are based on travel time and the turbulence parameters that are responsible for dispersion (Ramsdell, 1994). This approach, which has undergone extensive review, is considered to be more realistic than the approach used previously.

The horizontal dispersion parameter during the first hour following release is proportional to the product of a measurement of the horizontal component of the turbulence in the wind and the time since release. It is defined as follows:

$$\sigma_y(t) = 0.5 \int_0^t \sigma_v(t)dt, \tag{4-3}$$

where σ_v is the standard deviation of the fluctuations of the horizontal component of the wind vector perpendicular to the mean wind direction (m/s) at the puff center and t is the time since release (s). The dimensionless constant, 0.5, is the approximate value of a function similar to Equation 4-11 at t = 1,800 seconds, assuming a time scale of 1,000 seconds for horizontal turbulence (Irwin, 1983). As indicated, σ_v may vary along the puff path.

After the first hour, the rate of increase in the horizontal dispersion parameter is a function only of travel time. It is defined as:

$$\sigma_y(t) = \sigma_y(3600) + c_y(t - 3600), \tag{4-9}$$

where c_y is a constant. The default constant value is 0.2 m/s, which gives σ_y growth consistent with data compiled by Gifford (1982) for travel times in the 1- to 24-hour range.

The vertical dispersion parameter is a function of travel time, a measure of the vertical component of turbulence, and a function that accounts for decreasing effectiveness of turbulence in dispersion at long travel times. It is defined as:

$$\sigma_z(t) = \int_0^t \sigma_w(t)f_z(t)dt, \tag{4-10}$$

where σ_w is the standard deviation of the vertical component of the wind vector (m/s). Following Petersen and Lavdas (1986), RASCAL 4 uses two forms of the nondimensional function $f(z)$. For neutral and unstable atmospheric conditions (Pasquill-Gifford stability classes A–D), the function $f_z(\tau)$ is equal to 1.0. For stable conditions (Pasquill-Gifford stability classes E–G), the function is:

$$f_z(t) = \left[1.0 + 0.9(t/T)^{1/2}\right]^{-1}, \tag{4-11}$$

where T is a time-scale for turbulence that has a default value of 50 s. This function decreases the rate of growth of the vertical dispersion parameter from being proportional to travel time to the first power near the release point to being proportional to the square root of travel time after the first few minutes.

The turbulence parameters σ_v and σ_w are estimated as they are needed from atmospheric and surface conditions using relationships presented by Hanna et al. (1982) and Panofsky et al. (1977). For stable atmospheric conditions when the plume height z_p divided by the mixing height H is less than 0.9, the model uses the following expression:

$$\sigma_v = \sigma_w = 1.3u_*(1 - z_p/H), \tag{4-12}$$

where u_* is a scaling velocity in the atmosphere. (Section 6.4.2 discusses u_*, and Section 6.4.3 discusses the mixing layer thickness.) If the height of the plume is greater than $0.9\,H$, the following equation applies:

$$\sigma_y = \sigma_z = 0.13u_* \tag{4-13}$$

Throughout the mixing layer for neutral atmospheric conditions (Pasquill-Gifford stability class D), the model estimates the turbulence parameters using the following equation:

$$\sigma_v = \sigma_w = 1.3u_* exp\left(-2fz_p/u_*\right), \tag{4-14}$$

where f is the Coriolis parameter, which is a function of latitude and Earth's rotation rate. As described above, the lower limit for both σ_v and σ_w in neutral and stable conditions is $0.13u_*$.

For unstable conditions (Pasquill-Gifford stability classes A–C), the model estimates σ_v in the boundary layer using the following equation:

$$\sigma_v = u_*(12 - 0.5H/L)^{1/3}, \tag{4-15}$$

where L is the Monin-Obukov length. Two relationships are used to estimate σ_w in unstable conditions. In the lower half of the boundary layer, RASCAL 4 uses the relationship:

$$\sigma_w = 1.3u_*\left(1.0 - 3.0\,z_p/L\right)^{1/3} \tag{4-16}$$

In the upper half of the boundary layer, it uses the relationship:

$$\sigma_w = 1.3u_*(1.0 - 1.5\,H/L)^{1/3} \tag{4-17}$$

The second relationship follows from the first when the assuming that σ_w is independent of the height of the plume in the upper half of the boundary layer.

When the plume is above boundary layer, the model assumes that it is in a stable atmosphere. In this case, the model uses Equation 4-13 to estimate both σ_v and σ_w. Ultimately, the model uses a lower bound of 0.13 m/s for both σ_v and σ_w.

4.3.1 Low Wind Speed Corrections

Atmospheric dispersion experiments conducted in the vicinity of buildings (e.g., General Public Utilities Service Corporation, 1972; Start et al., 1978; Thuillier and Mancuso, 1980; Thuillier, 1988) indicate that the dispersion parameters used in RASCAL up to and including Version 3.0.5 tend to underestimate dispersion near buildings. Analysis of this dispersion data (Ramsdell and Fosmire, 1998) suggests that the apparent enhanced dispersion noted in the vicinity of buildings at low wind speeds is caused by underestimation of dispersion by the basic dispersion algorithms rather than by increased turbulence in the vicinity of buildings. Figure 4-1 shows this underestimation. This figure shows the variation of the ratio between values of χ/Q' estimated using the straight-line Gaussian model with dispersion parameters taken from the Pasquill-Gifford curves (Gifford, 1976) and the experimental values calculated from the observed data as a function of mean wind speed during the experiments.

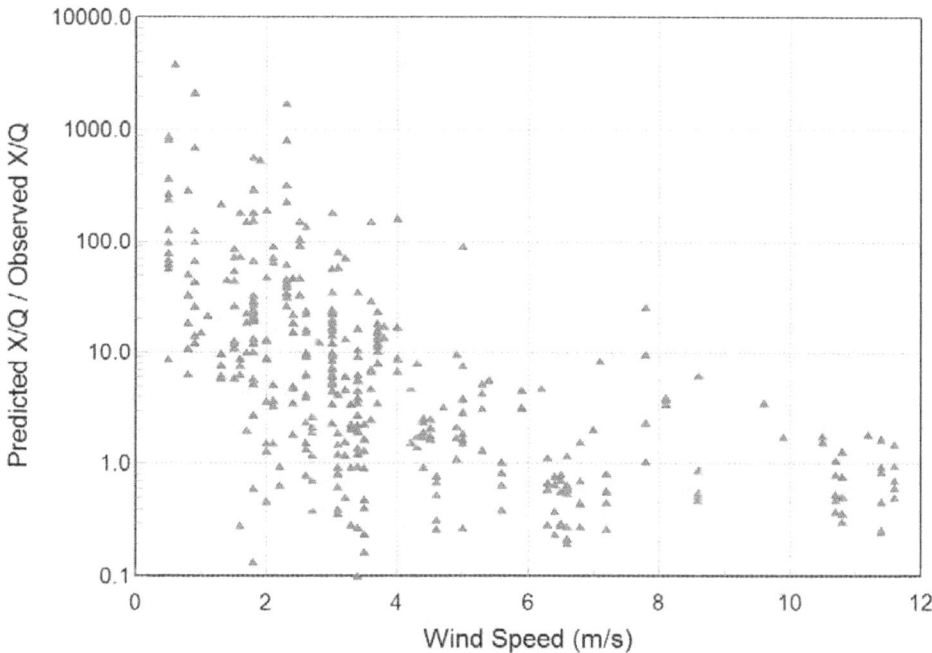

Figure 4-1   Ratios of predicted concentrations to concentrations observed in dispersion experiments as a function of wind speed

As in RASCAL 3, RASCAL 4 includes low wind speed corrections to the dispersion parameters to compensate for the low wind speed bias shown in Figure 4-1. The low wind speed corrections are applied through enhanced dispersion parameters. The enhanced dispersion parameters, Σ_y and Σ_z, are defined as:

$$\Sigma_y = \left(\sigma_y^2 + \Delta\sigma_y^2\right)^{1/2}, \text{ and} \tag{4-18}$$

$$\Sigma_z = (\sigma_z^2 + \Delta\sigma_z^2)^{1/2}, \tag{4-19}$$

where the enhancement terms $\Delta\sigma_y^2$ and $\Delta\sigma_z^2$ have the following form:

$$\Delta\sigma^2(t) = A(1 - (1 + t/T)exp(-t/T)), \tag{4-20}$$

where:

t = transport time (distance divided by the wind speed in TADPLUME and time since release in TADPUFF)
$A_y = (0.64T_y)^2$, where T_y is a horizontal time scale
$A_z = (0.845T_z)^2$, where T_z is a vertical time scale
$T_y = 1000$ s
$T_z = 100$ s

The constants in the definitions of A_y and A_z are dimensional (m/s). The constant values were determined from experimental data and the RASCAL 4 dispersion parameters using the optimization procedure described in Ramsdell and Fosmire (1998).

The enhancement terms are functions of wind speed and distance and are independent of stability and building dimensions. The enhancement terms increase with increasing distance from the release point until they reach an asymptotic limit that is a function of the time scales. The terms are large for low wind speeds and decrease as the wind speed increases. They are negligible for wind speeds above about 4 m/s. The correction terms also become negligible far downwind (large travel times) where they become small compared to the uncorrected dispersion parameter values.

4.3.2 TADPLUME Dispersion Parameters

The computational algorithms used for the standard Gaussian model in TADPLUME calculate dispersion parameters from integrated forms of Equations 4-8 and 4-10 with other parameters calculated as needed from Equations 4-11–4-20. The computational algorithms for the puff model used by TADPLUME during calm conditions calculate dispersion parameters from Equations 4-8 and 4-10 directly.

4.3.3 TADPUFF Dispersion Parameters

TADPUFF does a numerical approximation of Equations 4-8 and 4-10 as the puff moves along its trajectory. The time step in the integration varies depending on the size of the puff and the wind speed. For small puffs in moderate to high wind speeds, the time step is a few seconds. For large puffs, the time step is 5 minutes. For each step, the dispersion model parameters are updated using meteorological and surface conditions near the center of the puff.

4.3.4 Comparison of RASCAL 3 and RASCAL 4 Dispersion Parameters

The change in the approach for calculating dispersion estimates can result in significant changes in concentration and dose estimates. Figures 4-2–4-6 show various aspects of the changes in dispersion estimates as a function of wind speed and stability. The dispersion parameters in the figures include the low wind speed correction, and the mixing layer thickness limits the vertical dispersion parameters. The RASCAL meteorological data processor does not normally permit some wind speed/stability combinations shown (e.g., B or F stability with a 10-m/s wind speed). (Section 6.3.2 discusses normally permitted combinations.)

Figures 4-2 and 4-3 show the changes in atmospheric dispersion parameters for wind speeds of 1, 3, 5, and 10 m/s and B, D, and F stabilities. Figures 4-4 and 4-5 show the changes in ground-level concentration estimates. Figure 4-4 shows χ/Q as a function of distance, and Figure 4-5 shows the ratios between the χ/Q estimates from the two versions of the code. Finally, Figure 4-6 repeats the χ/Q versus distance curves shown for RASCAL 4 on the left side of Figure 4-4 to emphasize the variation in χ/Q with stability instead of wind speed.

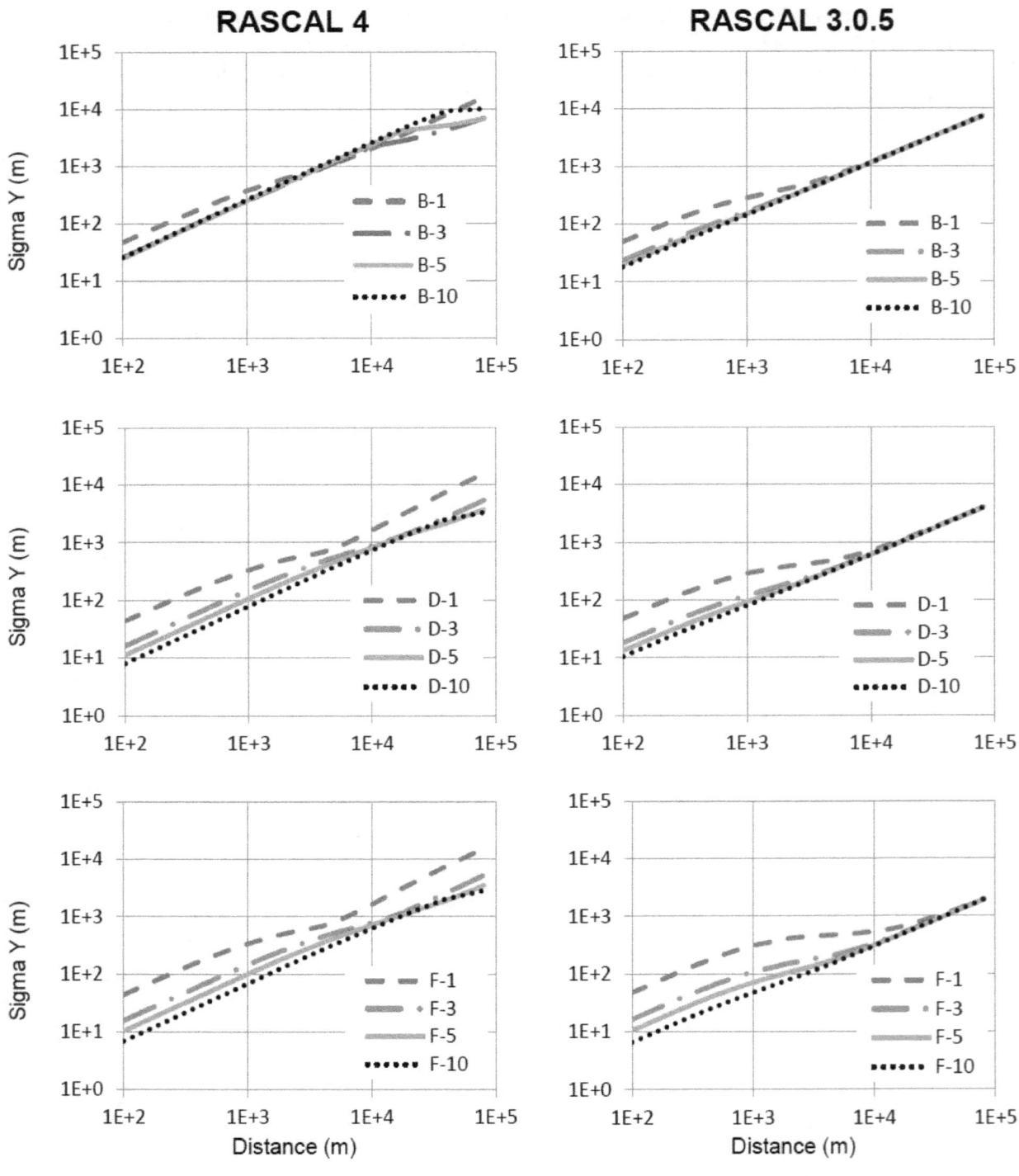

Sigma Y (m) versus Distance (m) for Stability Classes B, D, and F
with Wind Speeds of 1, 3, 5, and 10 m/s

**Figure 4-2   Comparison of horizontal dispersion parameters in RASCAL 4
to those in RASCAL 3.0.5 for various stabilities and wind speeds**

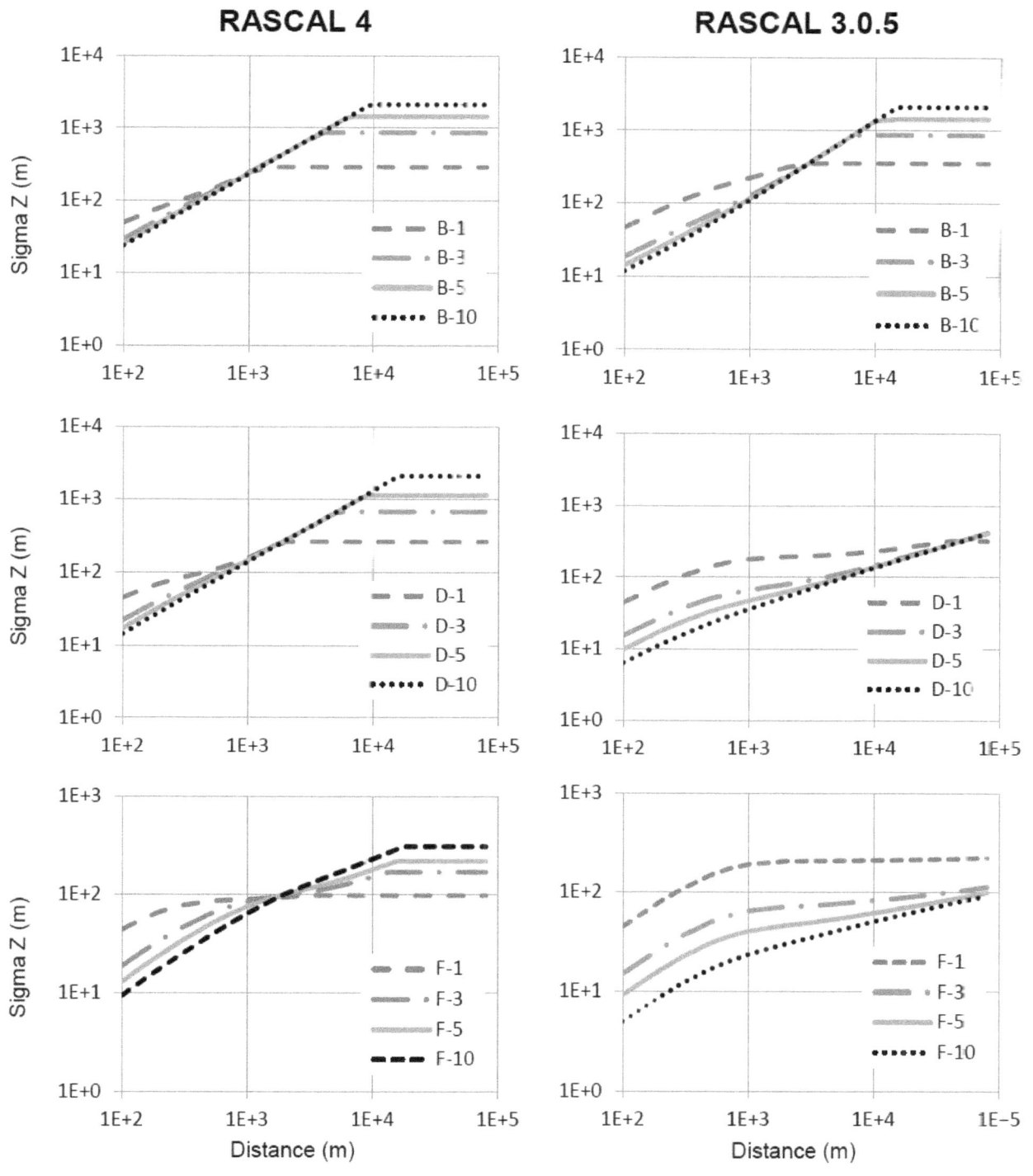

Sigma Z (m) versus Distance (m) for Stability Classes B, D, and F
with Wind Speeds of 1, 3, 5, and 10 m/s

**Figure 4-3   Comparison of vertical dispersion parameters in RASCAL 4
to those in RASCAL 3.0.5 for various stabilities and wind speeds**

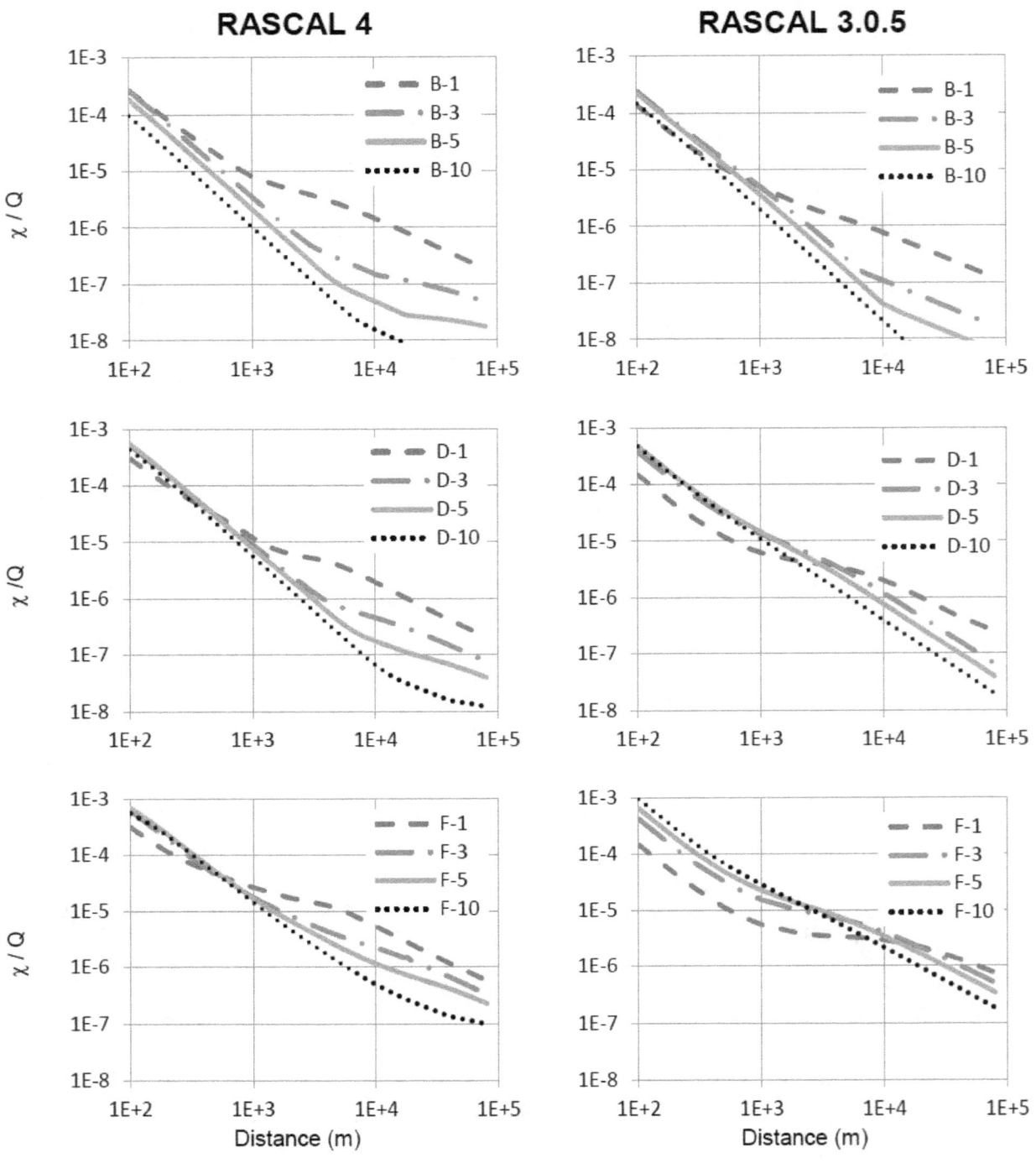

RASCAL 4   RASCAL 3.0.5

χ / Q versus Distance (m) for Stability Classes B, D, and F
with Wind Speeds of 1, 3, 5, and 10 m/s

**Figure 4-4   Comparison of RASCAL 4 χ/Q estimates for a ground-level release
to those of RASCAL 3.0.5 for various stabilities and wind speeds**

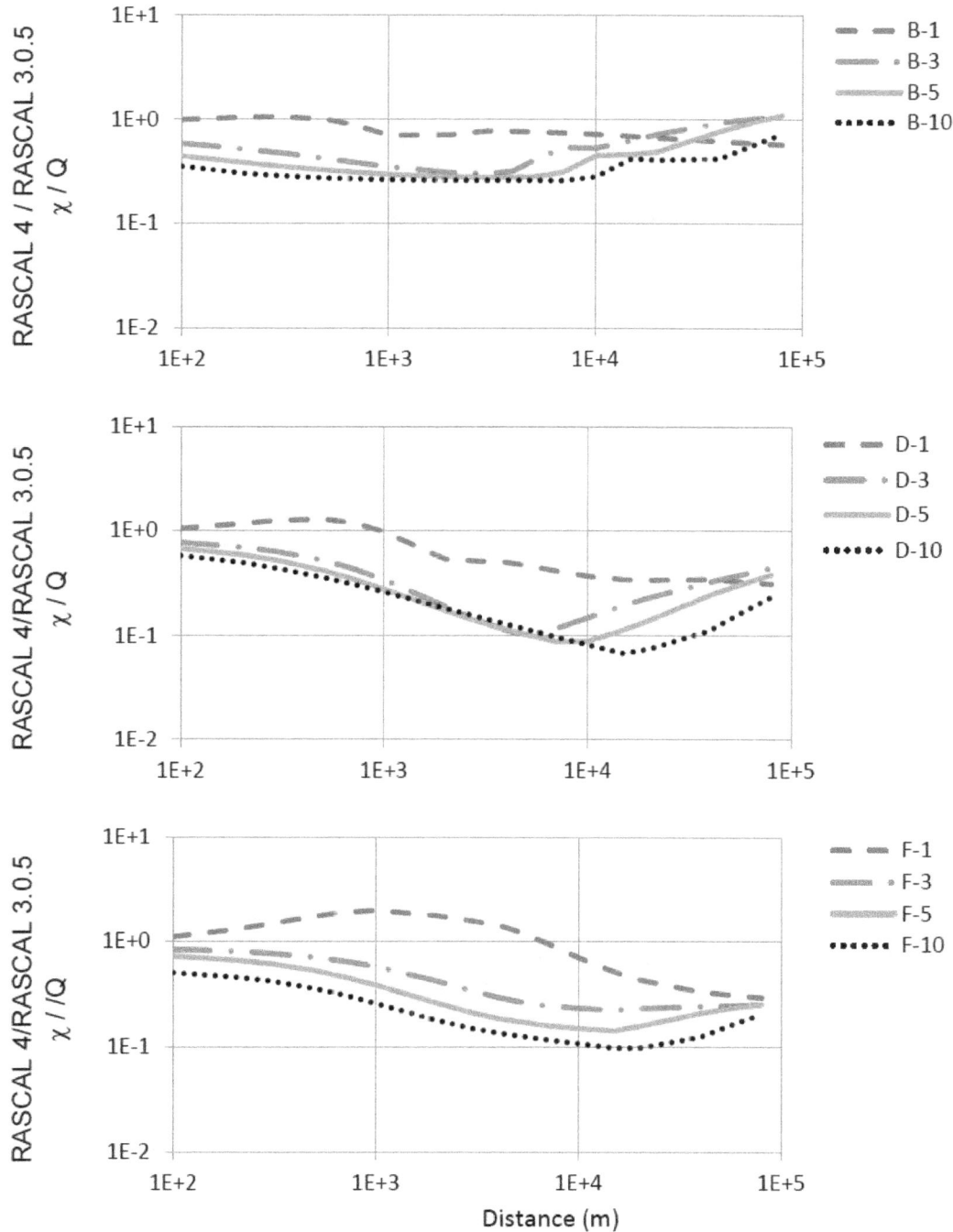

For Stability Classes B, D, and F with Wind Speeds of 1, 3, 5, and 10 m/s

**Figure 4-5 Ratio of RASCAL 4 to RASCAL 3.0.5 χ/Q for ground-level releases
for selected stabilities and wind speeds**

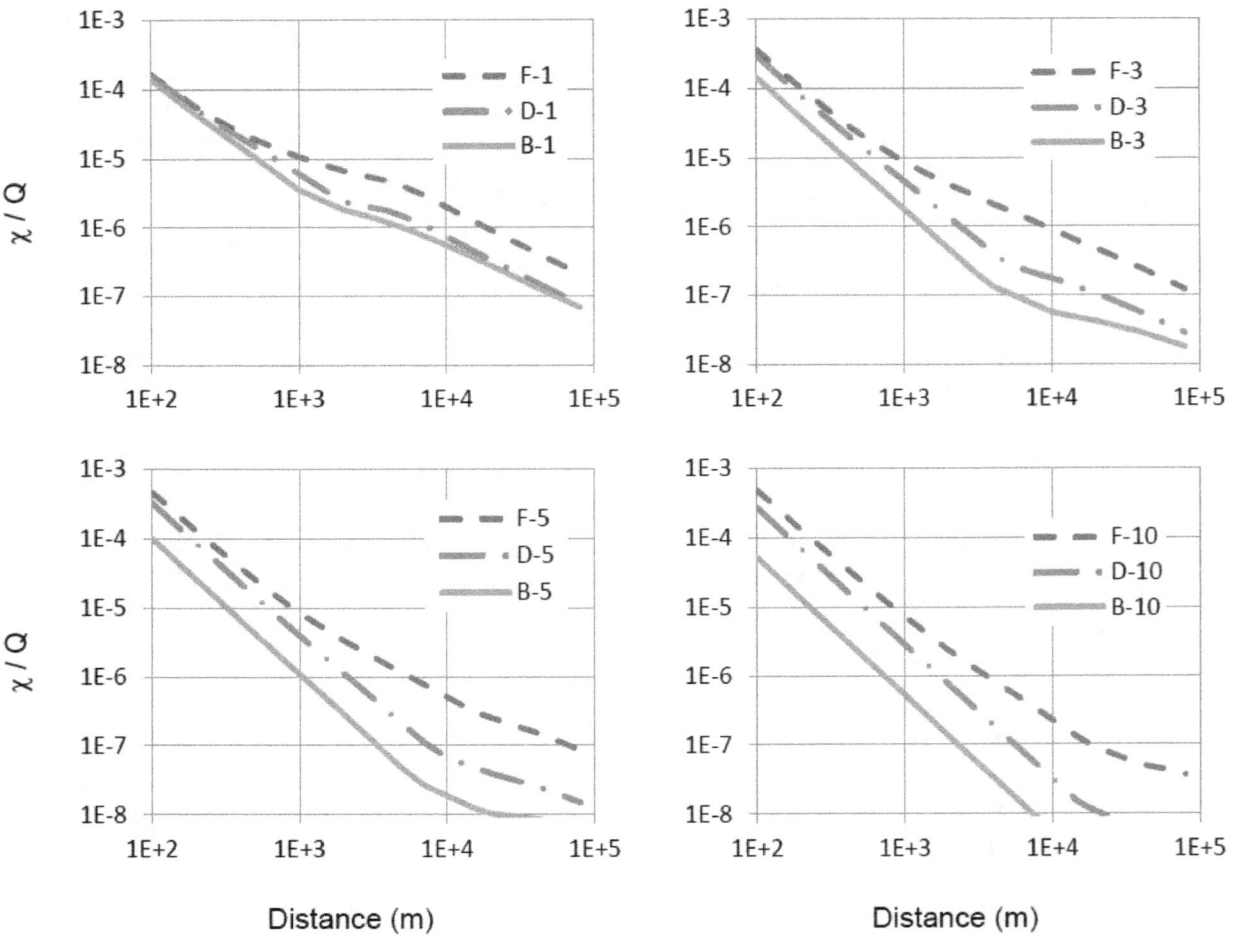

Comparisons for Stability Classes B, D, and F with Wind Speeds of 1, 3, 5, and 10 m/s

Figure 4-6   Sensitivity of RASCAL 4 χ/Q estimates for ground-level releases to stability for selected wind speeds

4.4 Mixing Layer Thickness

All TADPUFF and TADPLUME dispersion calculations include the mixing layer thickness. The thickness is passed to the atmospheric dispersion programs in the meteorological data files that are created by the meteorological data processing program. The meteorological data processing program has three options for determining the mixing layer thickness for each station: (1) the thickness may be estimated from meteorological data and surface roughness, (2) it may be estimated from climatological data, or (3) it may be entered directly. (Section 6.4.3 provides details on the estimation of mixing layer thickness.)

4.5 Deposition

RASCAL 4 calculates deposition for iodine and particles using the dry and wet deposition models in MESORAD (Scherpelz et al., 1986; Ramsdell et al., 1988). The activity deposited each time step is the product of the total deposition rate and the duration of the time step. At any time, the surface contamination (activity per square meter) is the sum of the activity deposited in the current time step plus the previously deposited activity. However, RASCAL 4 departs from the earlier versions in the algorithms used to calculate dry and wet deposition rates. As indicated at the beginning of this chapter, RASCAL 4 uses deposition algorithms developed and evaluated as part of the HEDR Project (Shipler et al., 1996; Ramsdell et al., 1996).

4.5.1 Dry Deposition

RASCAL 4 calculates dry deposition as the product of a deposition velocity and concentration. The product is:

$$\omega'_d = -v_{dd}\chi, \qquad (4\text{-}21)$$

where ω'_d is the deposition rate [(activity/m^2)/s], and v_{dd} is the dry deposition velocity (m/s). RASCAL 3.0.5 assumed a dry velocity of 0.003 m/s for iodine and for all particles. Deposition data summarized by Sehmel (1984) indicate that 0.003 m/s is a reasonable value for iodine under the assumption that about one-third of the iodine in the atmosphere is associated with particles, another one-third is in the form of reactive gases (e.g., molecular iodine (I_2) or hydrogen iodide), and the remaining one-third is in the form of nonreactive gases (e.g., methyl iodide (CH_3I)).

Current-generation-applied atmospheric models estimate the deposition velocity using an analogy with electrical systems. The analogy assumes that the deposition process is controlled by a network of resistances, and the deposition velocity is the reciprocal of the total resistance of the network. Resistances are associated with atmospheric conditions; physical and chemical characteristics of the material; and the physical, chemical, and biological properties of the surface. Seinfeld (1986) describes the resistance analogy.

Following the resistance analogy, the total resistance is made up of three components: (1) aerodynamic resistance, (2) surface-layer resistance, and (3) transfer resistance. These components are combined as follows:

$$v_{dd} = (r_a + r_s + r_t)^{-1}, \qquad (4\text{-}22)$$

where r_a, r_s, and r_t are the aerodynamic, surface, and transfer resistances (s/m), respectively. Equation 4-22 specifically applies to the deposition of gases. It may be extended to calculate deposition velocities for particles by adding a gravitational settling term. However, as a practical matter, the equation

4-17

may be used to estimate the deposition of fine particles (approximately 1 micron) because their settling velocity is small compared to r_a and r_s. The aerodynamic resistance is a function of wind speed, atmospheric stability, and surface roughness. RASCAL 4 estimates this resistance as:

$$r_a = u(10)/u_*^2, \qquad (4\text{-}23)$$

where $u(10)$ is the wind speed (m/s) at 10 m and u_* is the friction velocity. The friction velocity is a function of both wind speed and atmospheric stability. Consequently, the surface resistance is also a function of wind, atmospheric stability, and surface resistance. RASCAL 4 estimates this resistance as:

$$r_s = 2.6/(0.4 u_*), \qquad (4\text{-}24)$$

where 2.6 is a dimensionless empirical constant and 0.4 is von Karman's constant. The transfer resistance is associated with characteristics of the material and surface. However, RASCAL 4 uses the transfer resistance r_t as a means of placing a lower limit on the total resistance. Default values for this parameter are 10 s/m for reactive gases (e.g., I_2) and 100 s/m for fine particles (approximately 1 micron).

Noble gases (e.g., krypton) and gases that are nonreactive (e.g., CH_3I) are assumed to have deposition velocities of 0.0 m/s.

Deposition velocity in RASCAL 4 is no longer a constant because of the new deposition velocity algorithms. There are different deposition velocities for iodine and particles, and these deposition velocities change with changing meteorological conditions and surfaces. Table 4.1 lists deposition velocities for I_2 vapor and particles that RASCAL 4 calculated for representative stabilities and wind speeds using the new dry deposition velocity algorithms.

Table 4-1   RASCAL 4 Deposition Velocities* for Representative Stabilities and Wind Speeds

	STABILITY CLASS	WIND SPEED				
		1 m/s	2 m/s	3 m/s	5 m/s	10 m/s
Reactive Gases	B	0.0037	0.0062	0.0082	0.011	0.016
	D	0.0027	0.0047	0.0064	0.0091	0.014
	F	0.0021	0.0037	0.0051	0.0073	0.011
Particles	B	0.0047	0.0064	0.0073	0.0082	0.0090
	D	0.0039	0.0056	0.0065	0.0076	0.0086
	F	0.0031	0.0048	0.0058	0.0070	0.0082

*In meters per second.

4.5.2 Wet Deposition

Earlier versions of RASCAL calculated wet deposition of particles and gases using a washout model with washout coefficients that are only a function of precipitation type and a qualitative measure of intensity. RASCAL 4 treats wet deposition of particles and gases as two separate processes. Wet deposition of particles is calculated by using the washout model from earlier versions of RASCAL with coefficients that are functions of the precipitation type and rate. In contrast, wet deposition of gases is calculated by assuming that the concentration of gas in the precipitation is in equilibrium with the concentration in the air at ground level.

4.5.2.1 Particles

The wet deposition rate is calculated by using a washout model that assumes irreversible collection of particles as precipitation falls through the full vertical extent of the plume. In the washout model, the wet deposition rate for particles, ω'_w, is:

$$\omega'_w = -\lambda_p \int_0^\infty \chi dz, \tag{4-25}$$

where λ_p is a washout coefficient that is a function of the precipitation type; intensity; and, to a limited extent, temperature.

Calculation of washout coefficients in RASCAL is based on the discussion by Slinn (1984). For washout by rain, RASCAL 4 calculates the washout coefficient for particles as:

$$\lambda_p = (CEP_r)/\left(0.35P_n^{1/4}\right), \tag{4-26}$$

where:

C = an empirical constant with an assumed value of 0.5
E = the average collection efficiency assumed to be 1.0
P_r = the precipitation rate (mm/hr), and
P_n = the normalized precipitation rate (Pr/1mm/hr).

For snow, RASCAL 4 calculates the washout coefficient as:

$$\lambda_p = 0.2P_r \tag{4-27}$$

Hanna et al. (1982) and Slinn (1984) point out that the washout model is appropriate only for monodisperse aerosols and highly reactive gases.

4.5.2.2 Gases

In RASCAL 4, calculation of the wet deposition of gases is based on the assumption that the concentration of gases in the air and in the precipitation are in equilibrium. It follows from this assumption that the wet deposition rate of gases is proportional to the concentration of the gas in air near ground level and that wet deposition of gases can be modeled using a wet deposition velocity just as dry deposition of particles is modeled using a dry deposition velocity. Following Slinn (1984), RASCAL 4 estimates the wet deposition velocity for gases as:

$$v_{dw} = cSP_r, \tag{4-28}$$

4-19

where:

> v_{dw} = the wet deposition velocity
> c = a factor to convert the precipitation rate from millimeters per hour to meters per second
> S = a solubility coefficient
> P_r = the precipitation rate (water-equivalent rate for snow) (mm/hr)

The solubility coefficient for a gas is inversely related to the Henry's Law constant for the gas. Slinn (1984) provides guidance in selecting solubility coefficients. RASCAL 4 uses a default solubility coefficient of 1,000 for reactive gases (e.g., I_2). Solubility coefficients for nonreactive gases are about 3 orders of magnitude lower than those for reactive gases. Thus, wet deposition of nonreactive gases is extremely limited. RASCAL 4 neglects it.

When the temperature falls below about 3° C, the physical characteristics of snow change. As a result, there is a significant decrease in the wet deposition velocity of gases. RASCAL 4 does not calculate scavenging of gases by snow when the temperature is less than 3° C (i.e., the wet deposition velocity is set to zero).

Table 4-2 lists typical washout coefficients and wet deposition velocities. TADPLUME and TADPUFF convert washout coefficients to s^{-1} for use in the models.

Table 4-2   Typical RASCAL 4 Wet Depletion Parameters

PRECIPITATION TYPE	WASHOUT COEFFICIENT (h^{-1})	WET DEPOSITION VELOCITY (m/s)
Light rain	0.25	2.8×10^{-5}
Moderate rain	3.3	8.3×10^{-4}
Light snow	0.006	8.3×10^{-6}
Moderate snow	0.3	4.2×10^{-4}

4.5.3 Surface Contamination

The total deposition rate at any point is the sum of the dry and wet deposition rates at that point. This sum is:

$$\omega' = (v_{dd} + v_{dw})\chi + \lambda_p \int_0^\infty \chi \, dz \tag{4-29}$$

and, the total deposition C_{gi} of each isotope is:

$$C_{gi} = \int_0^t \omega_i \, dt \tag{4-30}$$

TADPUFF and TADPLUME calculate and store deposition by radionuclide for all radionuclides except noble gases. The codes assume that noble gases do not deposit and that those that are grown in following deposition are transported away from the point of deposition. The only noble gases that contribute to the doses are those that are included implicitly with a parent.

4.6 Depletion

Both atmospheric models in RASCAL 4 deplete the airborne activity to account for material deposited by dry and wet deposition.

In earlier versions of RASCAL, TADFLUME only depleted the plume to account for wet deposition TADPLUME now accounts for both wet and dry deposition. The model calculates the washout of particles by precipitation at each distance on the polar grid assuming an exponential decrease in activity with travel time (i.e., $Q'(x) = Q'_o \exp(-\lambda_p x/u)$). This approach is the same as before. The model estimates the activity lost because of dry deposition of particles and wet deposition of reactive gases by integrating the deposition flux (activity deposited per square meter per second) over the footprint of the plume between 100 meters and distances on the polar grid. The trapezoidal rule is used for the integration. TADPLUME does not deplete the plume for deposition between the source and 100 meters.

TADPUFF updates the activity in the puffs (Q) every 5 minutes. The model determines the activity to be removed from each puff by integrating the total deposition rate under the puff in space and time.

4.7 Decay and Ingrowth

TADPLUME and TADPUFF calculate decay and ingrowth using the algorithms and decay schemes described in detail in Appendix A. The decay schemes include as many as three daughters and account for branched decay chains.

Decay calculations are made at intervals determined by the code. For the default model parameters, the calculations are made at 5-minute intervals. The 5-minute decay and ingrowth calculations continue to the end of the calculation period specified by the user. If the duration of the calculation period specified by the user is less than 96 hours (4 days), TADPLUME and TADPUFF decay the material on the surface from the end of the calculation period to 96 hours in a single time step.

4.8 Iodine

Most radionuclides are treated as a single species, either particulate or gas. The isotopes of iodine are treated differently. Within the controlled containment environment following an accident, about 95 percent of the iodine is assumed to be in particulate form (cesium iodide), and the remaining 5 percent is assumed to be in gaseous form (I_2 or hydrogen iodide) (Soffer et al., 1995). However environmental data strongly suggest that once released to the environment, the distribution of iodine forms rapidly changes (Ramsdell et al.,1996).

Iodine is a special material that exists in three forms in the atmosphere. It is found in organic gases (e.g., CH_3I) and in inorganic gases (e.g., I_2), and it is attached to aerosol particles. RASCAL 4 treats iodine as a mixture of these three species with deposition characteristics that are a weighted average of the characteristics for each of the species. The weights assigned to each component are equal to the fraction of the total iodine in the component.

Experimental data on elemental iodine releases reported by Ludwick (1964) indicate that about two-thirds of the iodine changed form in the time required to travel 3.2 kilometers, about one-third was in organic species, and the remaining one-third was associated with particulate material. This partitioning of iodine is consistent with other results (e.g., in plumes from stacks at the Hanford Site (Ludwick, 1967; Perkins, 1963, 1964), in the plume following the Chernobyl reactor accident (Aoyama et al., 1986;

BIOMOVS, 1990; Bondietti and Brantley, 1986; Cambray et al., 1987; Mueck, 1988), and in natural atmospheric iodine (Voilleque, 1979). Consequently, RASCAL 4 assumes that the partitioning of iodine is independent of travel time.

The stochastic modeling in the HEDR Project (Ramsdell et al., 1996), based on the literature cited above, assumed that the iodine associated with particles was uniformly distributed between 5 percent and 45 percent, that the I_2 constituted 20 percent to 60 percent of the gaseous iodine, and that the remainder of the iodine was CH_3I. The range of values for each iodine fraction, based on these assumptions, is 0.05 to 0.45 for particles, 0.11 to 0.57 for I_2, and 0.22 to 0.76 for CH_3I. Note that the sum of the fractions is constrained to be 100 percent. Recent dose assessments (Apostoaei, 2005a; Apostoaei et al., 2005b; Thiessen et al., 2005) continue to base the treatment of iodine partitioning on the iodine partitioning done for the HEDR project (Ramsdell et al., 1996). RASCAL 4 is not a stochastic model. Therefore, it uses a partitioning based on the midpoints of the ranges for particles and I_2. The default partitioning in RASCAL 4 is 0.25 for particles, 0.30 for I_2, and 0.45 for CH_3I.

Using this partitioning, RASCAL 4 uses a weighted average dry deposition velocity for iodine. Table 4-3 lists the weighted average deposition velocities for representative wind speeds and stability classes. In addition, RASCAL 4 applies weightings based on partitioning to the washout coefficients and wet deposition velocities for iodine.

Table 4-3 RASCAL 4 Dry Deposition Velocities* for Iodines for Representative Stabilities and Wind Speeds

	STABILITY CLASS	WIND SPEED				
		1 m/s	2 m/s	3 m/s	5 m/s	10 m/s
Iodine	B	0.0023	0.0035	0.0043	0.0055	0.0072
	D	0.0018	0.0028	0.0035	0.0046	0.0063
	F	0.0014	0.0023	0.0030	0.0039	0.0055

*In meters per second.

The changes from RASCAL 3.0.5 to RASCAL 4 related to dispersion, deposition, and depletion of iodine are significant when considered individually. However, the net effect of these changes is not readily apparent from the equations because the changes have mixed effects on air concentrations and deposition. Figure 4-7 shows the net effect of the model changes for a 1-curie-per-second (Ci/s) release of iodine-131 (I-131) at ground level (10 meters) for three combinations of stability and wind speed. The top pair of graphs in Figure 4-7 is for B stability and a 10-m/s wind. The middle pair of graphs is for D stability and a 3-m/s wind. The bottom pair of graphs is for F stability and a 1-m/s wind.

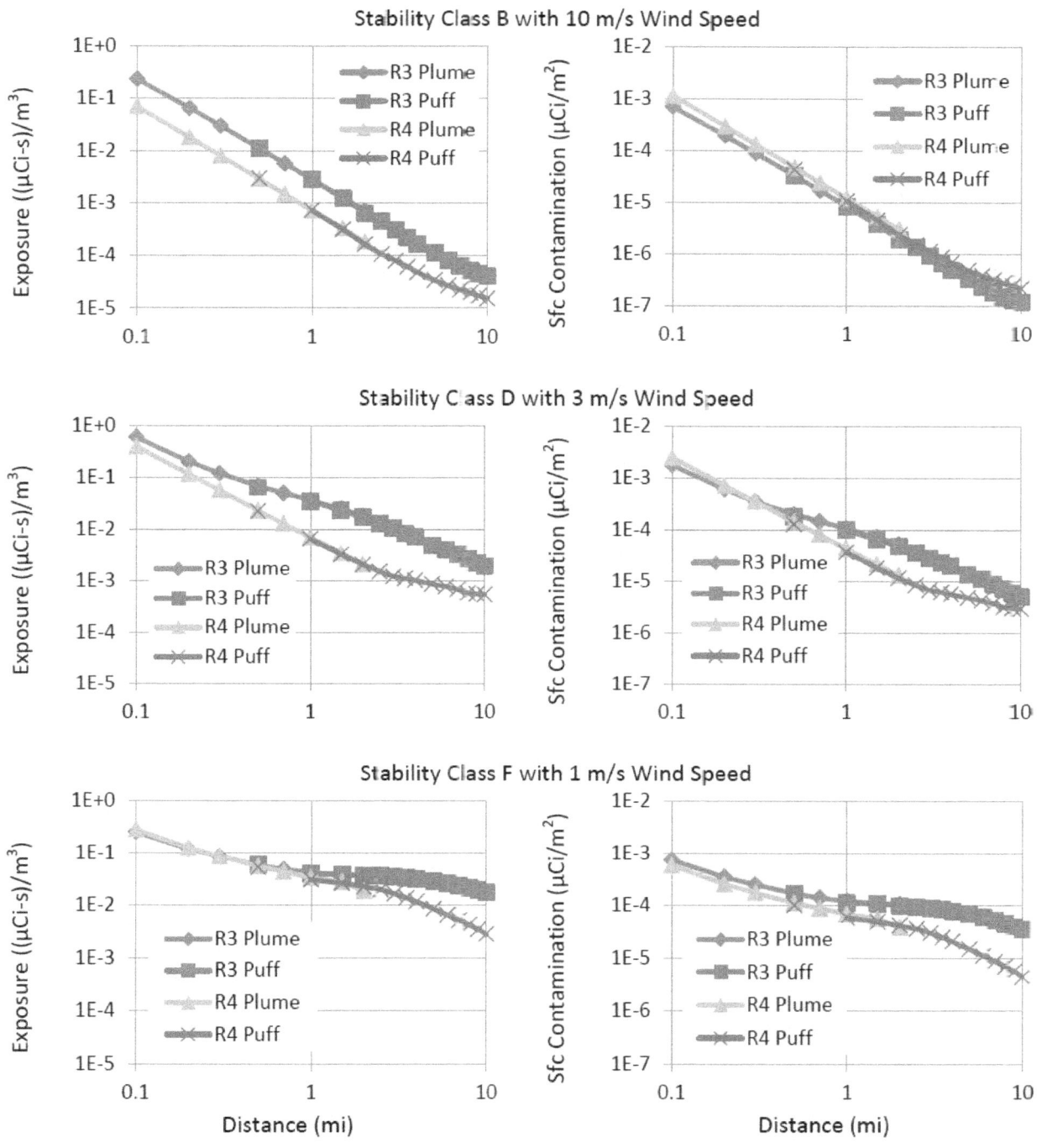

Figure 4-7 Comparison of I-131 exposure and surface contamination predicted by RASCAL 4 to those predicted by RASCAL 3.0.5

4.9 Dose Calculations

The RASCAL 4 calculation of dose commitments from inhaled activity (Section 4.9.1) and the calculation of groundshine doses (Section 4.9.2) are generally unchanged from RASCAL 3.0.5; however, the dose calculations can now treat temporal variations in the activity release rate and radionuclide mix. The cloudshine calculations have been revised to improve consistency between the TADPLUME and TADPUFF cloudshine dose estimates and to decrease TADPUFF computational time. The terms open- and closed-window exposure rates have been replaced by external gamma exposure rate and external gamma plus beta exposure rate. Section 4.9.3 discusses the cloudshine calculations, and Section 4.9.4 discusses the calculation of external gamma and beta exposure rates.

RASCAL 4.2 adds an option to use International Commission on Radiological Protection Publication 60 (ICRP-60), "1990 Recommendations of the International Commission on Radiological Protection," (ICRP, 1991), inhalation dose conversion factors (Eckerman and Leggett, 2002) from Federal Guidance Report No. (FGR)-13, "Cancer Risk Coefficients for Environmental Exposure to Radionuclides,". In addition, the code updates the ICRP-26 and ICRP-30 dose conversion factors from FGR-11, "Limiting Values of Radionuclide Intake and Air Concentration and Dose Conversion Factors for Inhalation, Submersion, and Ingestion," (Eckerman et al., 1988), and FGR-12, "External Exposure to Radionuclides in Air, Water, and Soil (Eckerman and Ryman, 1993). Finally, RASCAL 4.2 adds intermediate-phase dose calculations to TADPLUME and TADPUFF, which extend dose calculations to 50 years.

4.9.1 Organ Committed Dose Equivalents Caused by Inhalation

Both TADPLUME and TADPUFF calculate organ-committed dose equivalents and committed effective dose equivalents (CEDEs) for 15-minute periods. These dose equivalents are the sums over all radionuclides of products of the exposure to the radionuclide during the 15-minute period, a radionuclide- and organ-specific dose conversion factor, and the breathing rate. The general expression for the organ-committed dose equivalents is:

$$D_{15} = V_b \sum_n \left[DCF_n \int_t^{t+15} \chi_n(t)dt \right], \tag{4-31}$$

where:

D_{15} = organ-committed dose equivalent caused by inhalation during a 15-minute period
v_b = breathing rate
DCF_n = radionuclide n and the organ-specific dose conversion factor
χ_n = radionuclide n concentration
t = time

RASCAL 4 provides the two alternatives for calculating committed dose equivalent for the thyroid and the CEDE. Inhalation dose conversion factors from FGR-11 (Eckerman et al., 1988) and inhalation dose conversion factors from ICRP-60 (ICRP, 1997) are available in RASCAL 4.2. The inhalation dose conversion factors from FGR-11, which were used in RASCAL 4.1, have been revised using the 2008 updated ICRP-30 dose conversion factors included in DCFPAK2 (Eckerman et al., 2008). The ICRP-60 dose conversion factors included with RASCAL 4.2 are also from DCFPAK2. These factors are for adults (age of 7,300 days or 9,125 days). The ICRP-60 hydrogen-3 dose conversion factors are for tritiated water vapor (HTO). For consistency, the iodine dose conversion factors are weighted average dose conversion factors for a mixture of 25-percent particles, 30-percent reactive iodine gases, and 45-percent organic iodine gases. (Section 4.8 discusses the partitioning of iodine in RASCAL 4.) All of the remaining

ICRP-60 dose conversion factors are for inhaled particles assuming a particle size of 1 micron. For cases in which dose conversion factors are available for various clearance classes, RASCAL 4 uses the dose conversion factors for the clearance class that gives the highest doses.

Figures 4-8 and 4-9 compare the ICRP-26 and ICRP-60 CEDE and thyroid committed effective dose conversion factors for radionuclides released in a pressurized-water reactor loss of coolant accident. The figures identify the nuclides with the largest differences in dose conversion factors. In general, the differences in CEDE dose conversion factors are small compared to the uncertainties in the source term and dispersion. CEDE potential is defined as the sum over all nuclides of the product of total release (Bq) times the CEDE dose conversion factor (Sv/Bq). Changing from the FGR-11 dose conversion factors to the ICRP-60 dose conversion factors decreases the total CEDE dose potential by about 29 percent. Larger differences appear in the thyroid dose conversion factors. However, the nuclides with the largest differences account for less than 1 percent of the ICRP-60 thyroid dose potential. Overall, changing from FGR-11 to ICRP-60 dose conversion factors decreases the thyroid dose potential by about 3 percent.

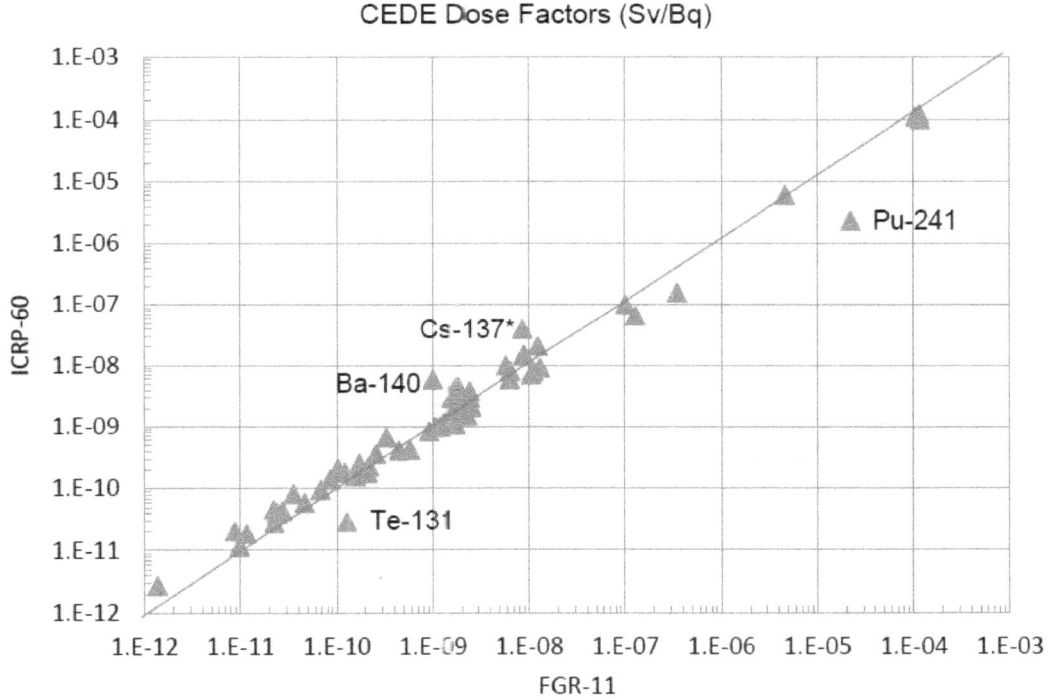

**Figure 4-8   Comparison of ICRP-26 and ICRP-60
CEDE dose conversion factors**

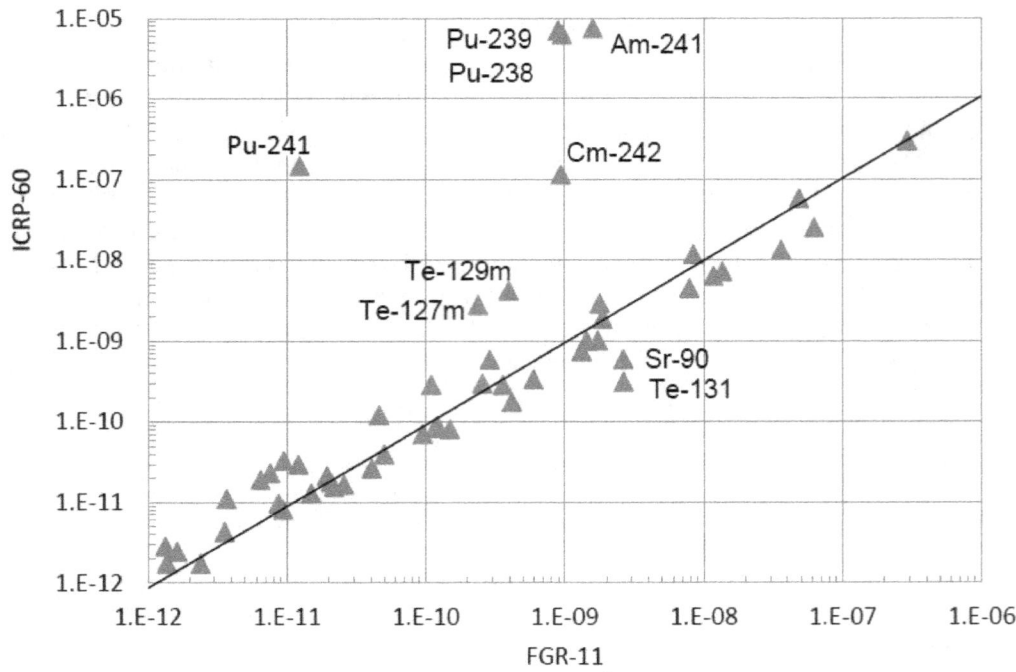

Figure 4-9   Comparison of ICRP-26 and ICRP-60 thyroid dose conversion factors

RASCAL 4 also includes acute bone, colon, and lung doses. These acute dose conversion factors are calculated using relative biological effectiveness (RBE) weighted dose conversion factors from the International Atomic Energy Agency (IAEA) report entitled, "Dangerous Quantities of Radioactive Materials, EPR-D-Values" (IAEA, 2006).

RASCAL 4.2 uses a breathing rate of 3.33×10^{-4} m^3/s for early-phase (first 96 hours) dose calculations. For intermediate-phase (96 hours to 50 years) dose calculations, RASCAL 4.2 uses a breathing rate of 2.57×10^{-4} m^3/s.

At the end of each 15-minute period, the committed dose equivalents at each receptor node are written to TADPLUME and TADPUFF output files. They are then set to zero before the start of the model calculations for the next 15-minute period.

4.9.2 Groundshine Doses

TADPLUME and TADPUFF calculate groundshine dose equivalents as the sum over all radionuclides of product of the surface contamination by the radionuclide and a radionuclide-specific dose conversion factor. The general expression for the groundshine dose equivalent is:

$$D_{gs} = SRF \times \sum_n \left[DCF_n \int_t^{t+15} C_{gn}(t)dt \right],$$
(4-32)

where:

D_{gs} = dose equivalent during the period

SRF = a surface roughness factor (0.82)
DCF_n = radionuclide n specific groundshine dose conversion factor
C_{gn} = radionuclide n surface concentration
t = time

The groundshine dose conversion factors are from Table III.3 in FGR-12 (Eckerman et al., 1993). Some of the dose conversion factors in Table III.3 have been revised. RASCAL 4.2 includes the revised dose conversion factors distributed with DCFPAK2 (Eckerman et al., 2008).

Note that the integration in Equation 4-32 is from t to $t+15$ minutes. RASCAL 4 uses these integration times because the surface concentration at any time is cumulative from the beginning of the event and is not set to zero at the beginning of the period. Once the surface becomes contaminated, the code will continue to calculate groundshine doses even if airborne material is not present.

RASCAL 4 also calculates a 4-day groundshine dose. This groundshine dose is the dose for the first 96 hours after the beginning of the release to the environment. The 4-day groundshine dose accounts for decay and ingrowth. It does not account for weathering that occurs after deposition because the weathering that occurs in the first 4 days after deposition is negligible (less than 0.3 percent).

4.9.3 Cloudshine Doses

Earlier versions of RASCAL calculated cloudshine doses using either a semi-infinite cloud model or a finite-plume model based on the model developed for MESORAD (Scherpelz et al., 1986). The first of these models assumes that activity is uniformly distributed through a large volume, and the second model assumes that activity is concentrated in a finite number of points distributed through a volume to represent the actual activity distribution. The use of the semi-infinite cloud model is usually inappropriate in the immediate vicinity of the release point, and application of the MESORAD finite-plume model to cloudshine dose estimates near the release point in TADPLUME was not satisfactory. Consequently, RASCAL 4 includes significantly revisions in the cloudshine dose models.

RASCAL 4 adds new cloudshine models based on line sources and plane sources that are analogous to the point-source model used in RASCAL 3.0.5. These models are used, along with the point-source model, until plumes and puffs grow to sufficient size to meet the assumptions associated with the semi-infinite cloud model. In addition, the point-source model which has been in use since MESORAD has been replaced by a faster model that uses precalculated dose rate versus distance curves. The radionuclide database provides the curves for each radionuclide for a 1 Ci point source and a 1Ci/m infinite-line source. The remainder of this section describes the RASCAL 4 cloudshine models.

4.9.3.1 TADPUFF Cloudshine Dose Calculations

The TADPUFF cloudshine dose calculations have three stages. Near the source where puff dimensions are small compared to the mean path length of photons, TADPUFF uses a revised version of the point-kernel dose model that previous versions of RASCAL used. When the puff radius becomes sufficient ($\sigma_y = 400$ m), an infinite-slab model is used to calculate cloudshine dose rates beneath the centerline of the plume. The model calculates the dose rate at ground level assuming that the activity in the plume is equally divided among 10 horizontal slabs with slab heights determined by the release height and vertical dispersion coefficients. The change in dose rate with distance from slabs is only because of buildup and absorption of photons; the change in dose rate across the plume is proportional to the crosswind variation of activity concentration in the slab. This model will be discussed further. Finally, when the vertical dimensions of the plume become sufficient ($\sigma_z = 400$ m),   a semi-infinite cloud model is used to calculate cloudshine.

Previous versions of RASCAL used the MESORAD finite-puff cloudshine model (Scherpelz et al., 1986; Ramsdell et al., 1988). This model first calculates composite characteristics (e.g., photon energies and photons per disintegration) of the gamma radiation from the radionuclide mix in a puff. Next, the model calculates the dose rate versus distance from a point source that has the composite characteristics using:

$$D'_p(\rho) = \frac{2.13 \times 10^6}{4\pi\rho^2} \sum_\gamma [f_\gamma \, B_\gamma(\mu_\gamma, \rho)e^{-\mu_{a\gamma}\rho}E_\gamma T_\gamma W_\gamma], \tag{4-33}$$

where:

$D_p'(\rho)$ = dose rate [(rem/h)/Ci]
ρ = distance from point source
f_γ = fraction of disintegrations producing photons of energy E_γ
$B_\gamma(\mu_\gamma,\rho)$ = buildup factor for air
$\mu_{a\gamma}$ = linear attenuation factor for air
E_γ = photon energy
T_γ = mass energy absorption coefficient for tissue ($\mu_{t\gamma}/\rho_t$)
W_γ = ratio of whole body dose to surface dose

The constant 2.13×10^6 is a collection of unit conversion constants to give dose rate in rem per hour per curie. The components of the constant are described following Equation 19 in Scherpelz et al. (1986). Previous versions of RASCAL used a constant value of 592, which gives dose rates in rem per second per curie.

The model then calculates the dose rate at ground level as a function of horizontal distance from the ground-level position of center of the puff. This calculation involves summation over volume elements distributed throughout the puff using:

$$D'(r) = \sum_i \sum_j \sum_k D'_p(\rho_{ijk})M_{ijk}, \tag{4-34}$$

where:

$D'(r)$ = dose rate at r [(rem/hr)/Ci]
r = distance from the receptor to the projection of the puff center on the ground
i,j,k = indices associated with the volume elements

$D'_p(\rho_{ijk})$ = dose rate at distance ρ from a point source in volume element ijk

ρ_{ijk} = distance from the center of the volume element ijk to the receptor

M_{ijk} = fraction of the total puff activity in volume element ijk

Cloudshine calculations assumed that puffs were circular cylinders with three layers. The volume elements were defined in 3, 5, or 8 annular rings with either 6 or 16 sectors. The fraction of activity in volume elements varies by annulus and level. Finally, accumulation of the dose at a receptor for a period is done by summing the product of dose rates for individual puffs and the time step duration over all puffs and all time steps in the period.

RASCAL 4 modifies this process by eliminating the calculation of composite characteristics from the gamma energies. In its place, TADPUFF calculates the dose rate versus distance from a point source that has all of the activity in the puff. Thus, the following equation replaces Equation 4-34:

$$D'_p(\rho) = \sum_{n=1}^{N} Q_n D'_{pn}(\rho),\tag{4-35}$$

where:

N = number of radionuclides

Q_n = activity of radionuclide n in the puff

$D'_{pn}(\rho)$ = dose rate at distance ρ from a 1-curie point source of radionuclide n

The overall puff geometry remains the same in TADPUFF as it was in earlier versions of RASCAL. However, the internal geometry has changed. TADPUFF divides the puff into 10 layers with each layer containing one-tenth of the activity. The number of annular rings has been fixed at 6 with each ring containing one-sixth of the puff activity, and the number of sectors has been fixed at 12. With these changes, the number of volume elements has been increased, and the fraction of activity in each volume element becomes one seven-hundred-twentieth of the total activity. In TADPUFF, Equation 4-35 becomes:

$$D'(r) = \frac{1}{720} \sum_i \sum_j \sum_k D'_p(\rho_{ijk})\tag{4-36}$$

As before, symmetry is used to reduce the computational load.

This geometry is assumed while the puff is small. When the horizontal dispersion parameter reaches 400 meters, the puff radius is large enough such that the horizontal variations in the cloudshine dose rate are directly proportional to the horizontal variation in concentration in the puff. At this point, TADPUFF changes from the point-source-based cloudshine model just described to a new plane-source-based model. The plane-source model also assumes that the puff is a vertical cylinder. However, instead of assuming that activity is distributed among volume elements, the plane-source model assumes that the activity is concentrated on 10 horizontal slabs (planes).

To calculate the activity in each slab, the plane-source model first integrates the concentration at the center of the puff ($y = 0$) vertically from the bottom of the puff to the top. This is similar to the integration done in calculating the depletion from wet deposition. It is:

$$\langle \chi \rangle_n = \int_{-\infty}^{\infty} \frac{Q_n}{(2\pi)^{3/2}\sigma_y^2\sigma_z}\exp\left[-0.5\left(\frac{r}{\sigma_y}\right)^2\right]\exp\left[-0.5\left(\frac{h_e}{\sigma_z}\right)^2\right]dz = \frac{Q_n}{2\pi\sigma_y^2}\exp\left[-0.5\left(\frac{r}{\sigma_y}\right)^2\right], \quad (4\text{-}37)$$

where $\langle \chi \rangle_n$ is the vertically integrated concentration of radionuclide n at the center of the puff. The plane-source model then divides this concentration by the number of slabs (10) to get the concentration in each slab.

Within the cylinder, the effective release height, the mixing layer thickness, and the vertical dispersion coefficient determine the vertical position of the slabs. When a Gaussian distribution is partitioned so that the area under the curve is divided into 10 equal parts and the center of mass of each part is determined, these centers of mass fall at $\pm0.127\sigma$, $\pm0.385\sigma$, $\pm0.674\sigma$, $\pm1.037\sigma$, and $\pm1.645\sigma$. Using this as a basis, the plane-source model initially estimates the slab heights as $h_e\pm0.127\sigma_z$, $h_e\pm0.385\sigma_z$, $h_e\pm0.674\sigma_z$, $h_e\pm1.037\sigma_z$, and $h_e\pm1.645\sigma_z$. The initial heights may lie below ground level or above the top of the mixing layer. The model adjusts any heights that fall outside these bounds to account for reflection by the boundaries. The model changes negative signs of heights to positive, and heights (h_p) that are above the mixing layer are replaced by $2H-h_p$.

The model calculates the dose rate at ground level from a slab as:

$$D'_{sl}(r,z) = \frac{0.1\sum_n Q_n DCF_{pn}}{2\pi\sigma_y^2}\exp\left[-0.5\left(\frac{r}{\sigma_y}\right)^2\right](1+k\mu z)\exp(-\mu z), \quad (4\text{-}38)$$

where:

> z = height of the slab above the receptor (m), which is assumed to be at 1 meter
> DCF_{pn} = dose conversion factor radionuclide n for an infinite plane [(rem/s)/(Ci/m^2)]
> μ = total gamma-ray absorption coefficient for air (m^{-1})
> k = ratio of energy in scattered photons to absorbed energy

RASCAL 4 calculates an approximate infinite-plane dose conversion factor as:

$$DF_{pn} = DCF_{sicn}/241.2, \quad (4\text{-}39)$$

where DCF_{sicn} is the semi-infinite cloud dose conversion factor [(rem/s)/(Ci/m^3)] and 241.2 is a constant with units of meters which was evaluated by comparing dose rates calculated by Equation 4-38 to semi-infinite cloud dose rates in plumes for which the semi-infinite cloud model is appropriate. Semi-infinite dose conversion factors contained in FGR-12 (Eckerman and Ryman, 1993) are used to estimate the infinite-plane dose conversion factors. These calculations were carried out for 30 radionuclides that are typically released in reactor accidents involving fuel damage. The standard deviation of the estimates of the constant value was 0.04.

In Equation 4-38, the term $(1+k\mu z)$ represents the buildup factor caused by scattered photons, and $\exp(-\mu z)$ represents the absorption of energy by the air. Healy and Baker (1968) and Healy (1984) discuss these terms. RASCAL 4 assumes that μ and k are constants with values appropriate for ~0.7 MeV photons ($\mu = 0.01$, $k = 1.4$) based on Figure 16.4 in Healy (1984).

Ultimately, RASCAL 4 calculates the dose rate at a receptor as:

$$D'_{sl}(r) = \frac{0.1 \sum_n Q_n DCF_{pn}}{2\pi\sigma_y^2} exp\left[-0.5\left(\frac{r}{\sigma_y}\right)^2\right]\sum_{i=1}^{10}(1 + k\mu z_i)\exp(-\mu z_i), \qquad (4\text{-}40)$$

where the summation is over all slabs.

When the vertical extent of the puff is sufficient for the semi-infinite cloud model to be appropriate ($\sigma_z > 400$ m, or a uniformly mixed plume with a vertical depth >600 m), the semi-infinite cloud model calculates the cloudshine dose rate as:

$$D'(r) = [\chi(r)/Q]\sum_n Q_n DCF_{sicn}, \qquad (4\text{-}41)$$

where $\chi(r)/Q$ is calculated using Equation 4-2 or a variation thereof, as appropriate. This calculation uses the semi-infinite cloud (air submersion) dose conversion factors from Table III.1 in FGR-12 (Eckerman and Ryman, 1993). Some of the dose conversion factors in Table III.1 have been revised. RASCAL 4.2 includes the revised dose conversion factors distributed with DCFPAK2 (Eckerman et al., 2008).

4.9.3.2 TADPLUME Cloudshine Dose Calculations

A new set of models has been developed for cloudshine dose calculations in TADPLUME. These calculations proceed in a manner similar to the calculations in TADPUFF. Near the source, the plume is divided into a large number of equal-strength line sources spaced to properly represent the distribution of activity in the plume. TADPUME uses the dose rates from these line sources to calculate the ground-level dose rate as a function of horizontal distance from the plume axis. It then uses this relationship to calculate dose rates and 15-minute doses at receptor locations. When the width of the plume is sufficient ($\sigma_y = 400$ m), TADPLUME switches from the line-source model to an infinite-slab model. Finally, when the vertical dimensions of the plume are sufficient to make the semi-infinite cloud model appropriate ($\sigma_z = 400$ m or a uniformly mixed plume with a thickness of 600 meters), TADPLUME switches to a semi-infinite cloud model.

RASCAL 4 calculates finite-plume cloudshine dose rates from line source as:

$$D'_l(\rho) = \sum_{n=1}^{N} Q'_{ln}D'_n(\rho), \qquad (4\text{-}42)$$

where:

> $D'_l(\rho)$ = dose rate (rem/s) at a distance ρ from an infinite line source of Q'_{ln} (Ci/m)
> Q'_{ln} = line-source strength (Ci/m), $Q'_{ll} = Q'_n/u$ for a Q'_n (Ci/s) point source
> $D'_n(\rho)$ = line-source dose rate conversion factor [(rem/s)/(Ci/m)] for radionuclide n

Equation 4-42 is analogous to Equation 4-38 with changes in the definitions of source term and dose conversion factors.

TADPLUME combines the line-source dose rates to obtain the plume dose rate by summing over all line sources just as the point-source dose rates were combined to get a puff dose rate. The horizontal dispersion parameter σ_y determines the number of lines used. If $\sigma_y > 200$ m, 100 lines (10×10) are used to describe the concentration distribution in the plume. Otherwise, 36 lines (6×6) describe the concentration distribution. In either case, the lines are spaced horizontally and vertically so that each line represents the same fraction of the total activity in the plume. Finally, RASCAL4 calculates the plume dose rate as:

$$D'(y) = \frac{C_R}{N_l} \sum_i \sum_j D'_l(y_{ij}),$$

(4-43)

where:

$D'(y)$ = plume dose rate (rem/s)
y = distance from the ground-level projection of the center of the plume (m)
C_R = finite-line correction factor
N_l = number of line sources (36 or 100)
i,j = line-source indices
$D'_l(y_{ij})$ = infinite-line line-source dose rate

Equation 4-43 includes a finite-line source correction factor to account for the fact that the plume does not extend upwind of the release point. A correction factor could be calculated by numerical integration of a rather complex equation. However, the following equation can be used to estimate an approximate correction factor of adequate accuracy for emergency response dose calculations:

$$C_R = 0.5 \left[1 + \frac{x}{(R^2 + h_e^2)^{1/2}} \right],$$

(4-44)

where:

x = downwind distance (m) to a point beneath the plume centerline at the intersection of the plume centerline and a perpendicular line passing through the receptor
R = distance (m) from the release point to the receptor
h_e = effective release height (m)

In the case of a ground-level release and a receptor on the plume centerline, the correction factor will be 1. However, generally the correction factor is less than 1. For a 0.7 MeV photon, the correction factor given by Equation 4-44 corresponds to a receptor at a position approximately 100 meters off of the plume centerline. Doses will be slightly overestimated for receptors that are closer than 100 meters and slightly underestimated for receptors that are farther than 100 meters from the centerline.

When the horizontal dispersion parameter exceeds 400 meters, TADPLUME shifts from a line-source-based finite-plume model to an infinite-plane model. The infinite-plane model used by TADPLUME is similar to the model used by TADPUFF. The differences between the two models are associated with the calculation of concentrations, not with the cloudshine calculation. Thus, for TADPlume, Equation 4-40 for TADPUFF becomes:

$$D'_{sl}(r) = \frac{0.1 \sum_n Q'_n DF_{pn}}{2\pi \sigma_y u} exp\left[-0.5 \left(\frac{r}{\sigma_y} \right)^2 \right] \sum_{i=1}^{10} (1 + k\mu z_i) exp(-\mu z_i)$$

(4-45)

Finally, when the vertical extent of the plume is sufficient (σ_z = 400 m, or a uniformly mixed plume with a 600 m vertical extent), TADPLUME shifts to a semi-infinite-plume cloudshine model. Similarly, Equation 4-41 for TADPUFF becomes:

$$D'(r) = [\chi(r)/Q'] \sum_n Q'_n DCF_{sicn}, \tag{4-46}$$

where $\chi(r)/Q'$ is calculated using Equation 4-5 or a variation thereof, as appropriate.

4.9.4 Gamma and Gamma Plus Beta Exposure Rates

RASCAL 4 calculates a gamma exposure rate in mrad per hour for use in comparisons to field radiation measurements. It calculates this exposure rate from cloudshine and groundshine dose rates with adjustment for the difference in energy absorption coefficients of tissue and air. RASCAL 4 calculates the gamma exposure rate:

$$D'_{gair} = \frac{(D'_{cs} + D'_{cg})}{KD}, \tag{4-47}$$

where:

D'_{gair} = gamma exposure rate in air
D'_{cs} = effective dose equivalent (EDE) from cloudshine
D'_{gs} = EDE from groundshine
KD = conversion factor for kerma to organ dose assumed to have a value of 0.7 (Cember, 1996)

Gamma exposure rates are available for both the close-in grid and the Cartesian grid. The RASCAL 4 maximum value table does not include them, but they are found under Detailed Results in the RASCAL 4 output. It is important to note that these are exposure rates that reflect both where the plume is and where the plume has been. Consequently, there may be a significant difference in gamma exposure rates on the two grids. The plume model used for the close-in grid instantaneously transports material from the release point to the receptor, whereas the puff model transports material using the mean wind speed. Figure 4-10 illustrates potential differences in gamma exposure rates as a function of distance from the source and time since the beginning of release for a typical reactor accident in a steady 4-mile-per-hour wind. The top panel compares exposure rates for different time periods as a function of distance. The lower panel compares the exposure rates for different distances as a function of time. The difference increases with increasing distance and decreases with increasing time since the beginning the release. When the gamma exposure rates from the two models differ significantly, the gamma exposure rate from the Cartesian grid (puff model) is likely to be more realistic.

RASCAL 4 calculates the gamma + beta exposure rate as the skin dose rate assuming the appropriate correction factor for adjusting skin dose rate to beta exposure rate is 1.0. The initial estimate of gamma + beta exposure rate based on the skin dose rate is compared with the gamma exposure rate. If the gamma exposure rate is larger than the initial estimate of gamma + beta exposure rate, the gamma exposure rate is used as the gamma + beta exposure rate. Like the gamma exposure rate, the gamma + beta exposure rate is only found under Detailed Results in the RASCAL 4 output.

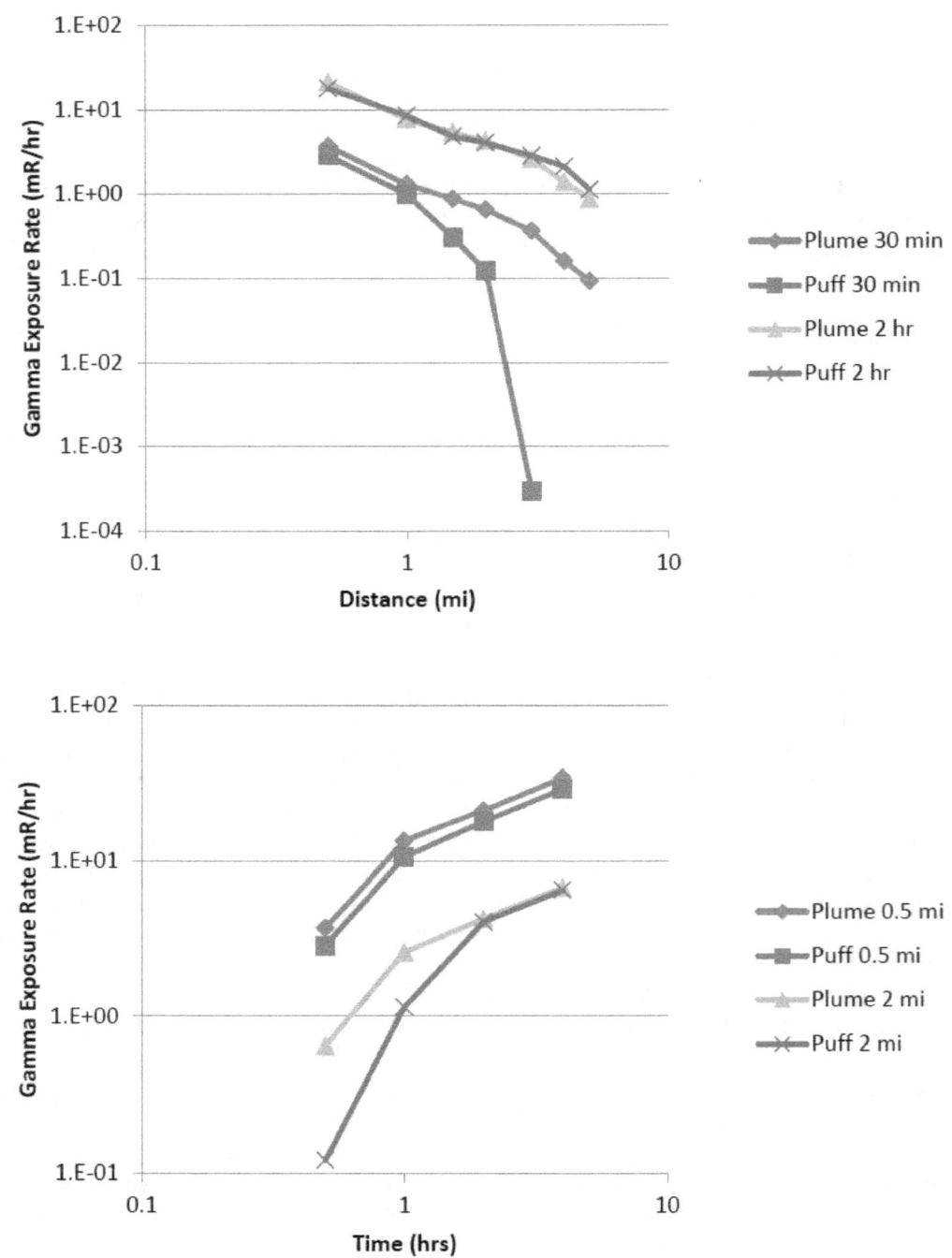

Figure 4-10 Comparisons of gamma-ray exposure rates calculated by the plume (close-in grid) and puff models (Cartesian grid)

4.9.5 Total Effective Dose Equivalent

The early-phase (plume-phase) total effective dose equivalent (TEDE) that RASCAL 4 calculates is the sum of the external gamma dose (cloudshine) from the plume, the CEDE, and the external dose over a 4-day period from radionuclides deposited on the ground (4-day groundshine dose). The calculations for this TEDE assume that no protective actions, such as evacuation or sheltering, are taken. Thus, the calculations assume that people are outdoors during plume passage and will remain outdoors exposed to groundshine from deposited radionuclides for 4 days after deposition of the radionuclides.

Thus, the early-phase TEDE that RASCAL 4 calculates is larger than the TEDE that would be expected for people who took protective actions or who continued their normal activities (spending much time indoors).

RASCAL 4 calculates dose under the assumption that no actions to reduce dose are taken to determine whether doses without any protective actions would exceed the Environmental Protection Agency (EPA) protective action guides (EPA, 1992). The need for protective actions is based on the TEDE that a person would receive if no protective actions of any type were taken, including actions such as simply spending some time indoors.

Figure 4-11 compares the results of RASCAL 4 TEDE calculations for a core-uncovered accident during D stability, 3 m/s wind conditions with the results of RASCAL 3.0.5 calculations for the same accident. The upper panel of the figure shows the comparison for a ground-level (10-meter) release, whereas the lower panel shows the results for a release at 100 meters. The decrease in dose from RASCAL 3.0.5 to RASCAL 4 is consistent with the χ/Q ratio shown in Figure 4-5 and with the relative magnitudes of the dispersion parameters shown in Figures 4-2 and 4-3. Figure 4-12 compares the TEDE components for this case. The larger vertical dispersion parameter in RASCAL 4 clearly causes the close-in increase in TEDE by bringing the plume to the ground sooner than would take place in RASCAL 3.0.5. Note that there is little difference in the cloudshine doses from RASCAL 3.0.5 to RASCAL 4.

The RASCAL 4 dose estimates should not be used as an estimate of the TEDE that people who did not intentionally take protective actions would receive because even normal everyday activities will reduce doses to below those estimated by RASCAL 4.

RASCAL 4 could provide a more realistic estimate of the doses that people would actually receive, but the estimate would require some effort. To account for evacuation, the end of calculation time can be set to the time at which people evacuate. The TEDE then is the sum of the inhalation dose, cloudshine dose, and period groundshine dose. The dose components for use in calculating TEDE doses with protective actions are found in the Detailed Results portion of the RASCAL 4 output. Protective actions may include sheltering before evacuation. To account for sheltering before evacuation, each dose component must be reduced by an appropriate reduction factor before summation of the three dose components.

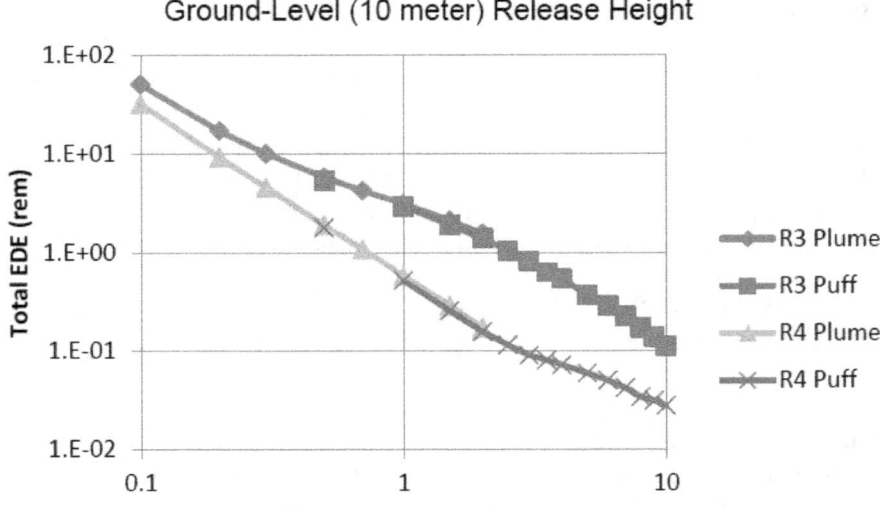

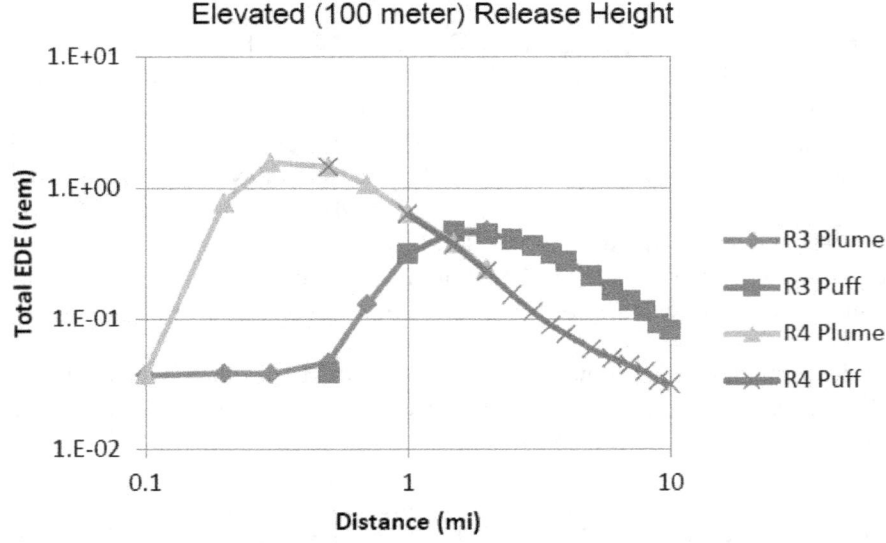

Figure 4-11   Comparison of TEDE calculated by RASCAL 4 to TEDE calculated by RASCAL 3.0.5

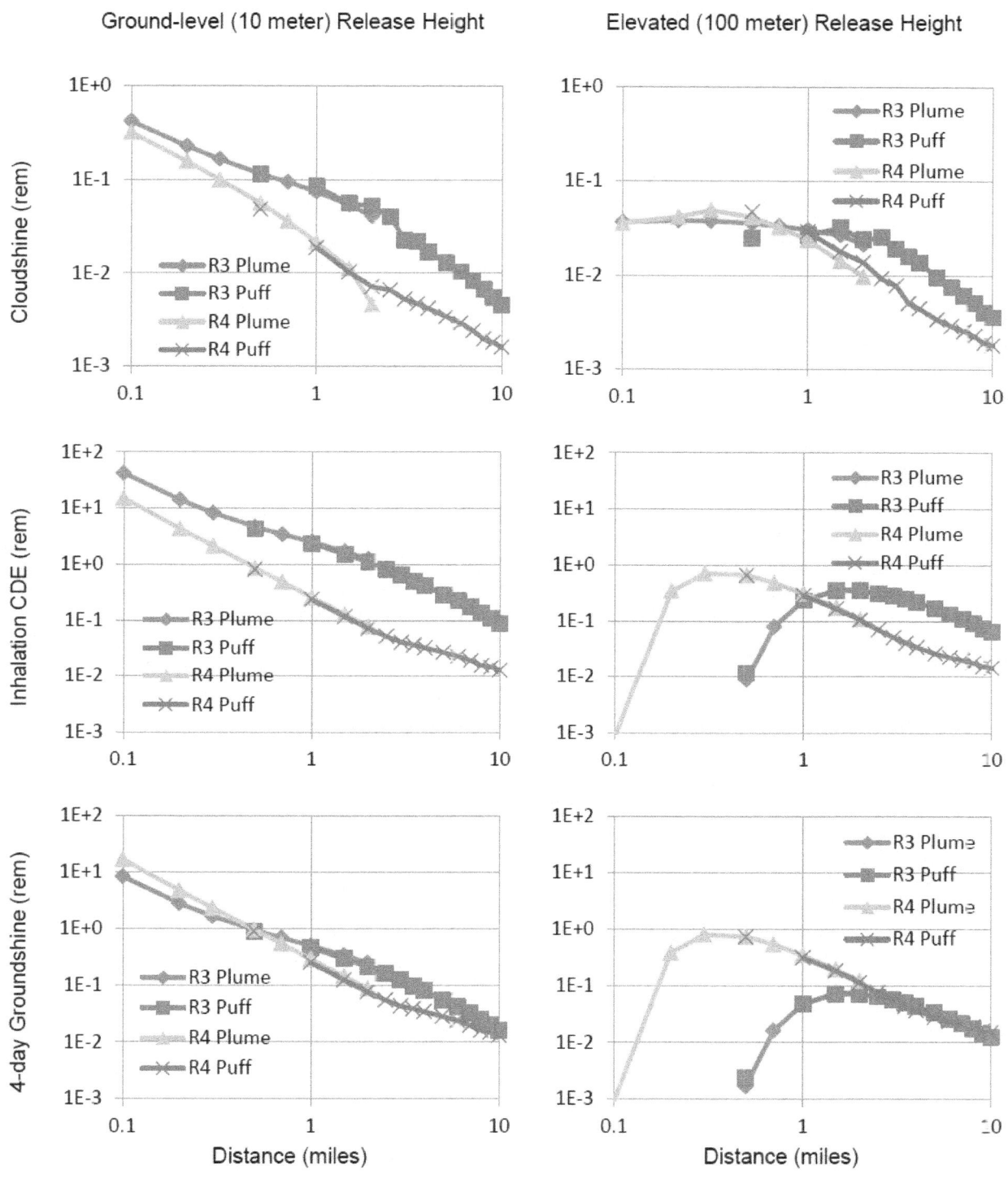

Figure 4-12   Comparison of RASCAL 4 and RASCAL 3.0.5 TEDE component doses

4.9.6 Acute Doses

The doses that RASCAL 4 reports are generally based on 50-year dose commitments. RASCAL 4 also calculates acute organ doses that can be used to determine whether acute (early or deterministic) health effects may occur. The calculated acute doses to red bone marrow, colon, and lung can be found under the Detailed Results section of the RASCAL 4 results. The Detailed Results include a total acute bone dose. The total acute bone dose is the sum of the cloudshine dose, the period groundshine dose, and the acute inhalation dose to the red bone marrow. The acute inhalation dose conversion factors in RASCAL 4 (IAEA, 2006) give full weight to doses delivered to the organ early on and less weight to doses delivered at later times because doses delivered late do not contribute to early-onset health effects.

The IAEA report (2006) does not provide acute dose conversion factors for all radionuclides in the RASCAL 4 database. The report does not include some radionuclides because they have short half-lives and it does not include other radionuclides because they have little importance for radiological emergency response.

4.9.7 Intermediate-Phase Doses

RASCAL 4 calculates intermediate-phase doses for comparison to the EPA's intermediate-phase protective action guides (EPA, 1992) in both the source term to dose and field measurement to dose modules. The computational methods in the two RASCAL 4.2 modules are identical. Specifically, RASCAL 4 calculates the intermediate-phase doses for the first year after a release, for the second year after a release, and for 50 years after a release. The intermediate-phase doses are based on ground contamination at the end of the early phase. In the source term to dose module, the maximum value table presents the maximum values for the first and second year intermediate-phase doses. The Detailed Results in the RASCAL 4 output presents all of the intermediate-phase doses. These doses account for decay, ingrowth, weathering, and resuspension. Chapter 7 discusses the calculation of intermediate-phase doses in detail.

Intermediate-phase doses calculated by the source term to dose and field measurement to dose modules are the same for a given mix of radionuclides. For most radionuclides, the intermediate-phase doses calculated by RASCAL 4 are identical to those calculated by TurboFRMAC 2011 (SNL, 2009; SNL, 2010). The differences in the intermediate-phase doses for iodine are related to differences in the representation of iodines in the two codes. RASCAL 4 treats iodines as a mixture of particle and gaseous forms, whereas TurboFRMAC 2011 treats iodines as all particles. In addition, some differences in the intermediate-phase doses are because of differences in representation of radionuclide decay chains. These differences are generally small.

4.10 Numerical Artifacts

Two numerical artifacts appear in the maximum value table when RASCAL 4 is run with constant, spatially uniform meteorology. These artifacts are the result of the discrete spacing of the Cartesian grid and finite time steps used in TADPUFF. The first of these artifacts appears in the Maximum Value table when the wind direction is not from a cardinal direction (i.e., not from north, east, south, or west). The second numerical artifact appears in the finite-plume cloudshine calculations near the release point when puff dimensions are small compared to the spacing between nodes on the Cartesian grid.

Figure 4-13 illustrates the first numerical artifact. The solid line in the figure is based on TEDE values for a reactor accident taken from the RASCAL 4 close-in maximum value table. These values are independent of wind direction. The dashed line is based on the TEDE values for the same accident assuming a cardinal wind direction (north, east, south, or west). The difference between the lines is the result of the time-dependent nature of the puff model (i.e., the travel time from the release point to the Cartesian grid node (receptor)). The artifact is that the doses for a noncardinal wind direction (e.g., northeast and east-southeast) are a function of wind direction. The dots show the TEDE values for a wind direction 45 degrees off of a cardinal heading. These points show the same trend as the values for the cardinal heading, but they do not fall on the line. The cause of the artifact is that doses for cardinal headings are calculated at plume centerline at exactly the distances shown; for off-cardinal wind directions, the doses shown in the table are maximum values for nodes closest to the distances shown. In general, the nodes are neither at the plume centerline or at exactly the distances shown. Locations of the points for which the doses are calculated may be determined from the numerical table in the Detail Results portion of the RASCAL 4 output. This artifact is generally not apparent when real meteorological conditions are used in the calculation because of wind direction variability, and it can be avoided by specifying a cardinal wind direction if the wind direction is constant.

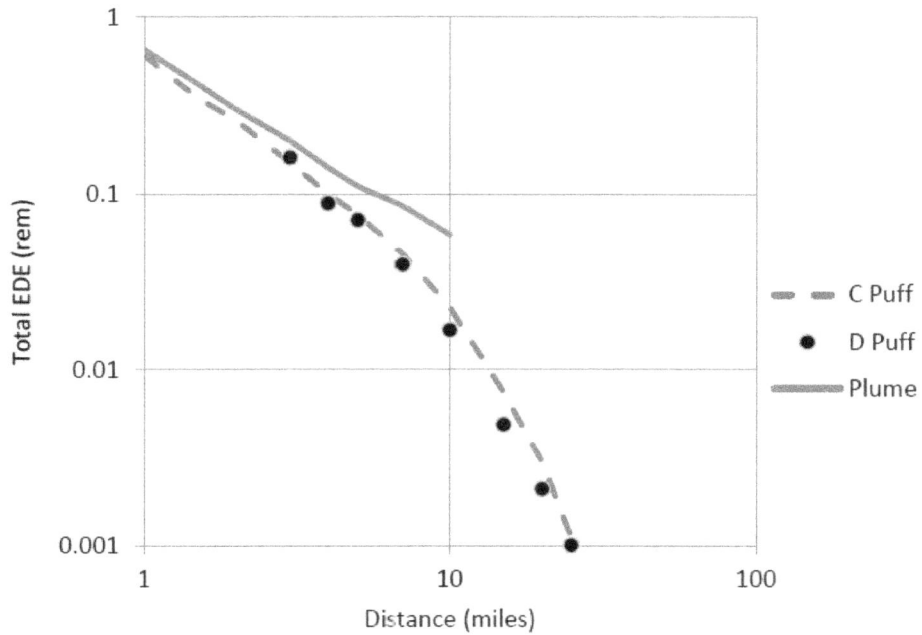

Figure 4-13   Numerical artifact in the puff model maximum value table associated with off-cardinal wind directions

Figure 4-14 illustrates the second numerical artifact by the small-scale variations in the cloudshine dose curves. The top panel shows RASCAL 3.0.5 results, and the bottom panel shows RASCAL 4 results. The cause of the artifact is related to puff size and the discrete steps used to move the puff. Near the release point where the puff (plume) dimensions are small, the code assumes that the radionuclides are concentrated at a relatively small number of points; it calculates the cloudshine doses at nodes by summing the contribution from each point and moves the puffs in discrete steps. As a result, under constant meteorological conditions, some nodes are consistently closer to the puff center and thus receive higher doses than other nodes. The longer the constant meteorological conditions persist, the more noticeable the difference between consecutive nodes becomes. Again, this artifact is generally not apparent under real meteorological conditions in which wind directions, wind speed, and atmospheric conditions vary in time. RASCAL 4 randomly varies the speed of movement of each puff ±15 percent from the nominal speed as the puff moves downwind to minimize this artifact. A comparison of the RASCAL 3.0.5 and RASCAL 4 panels shows this improvement.

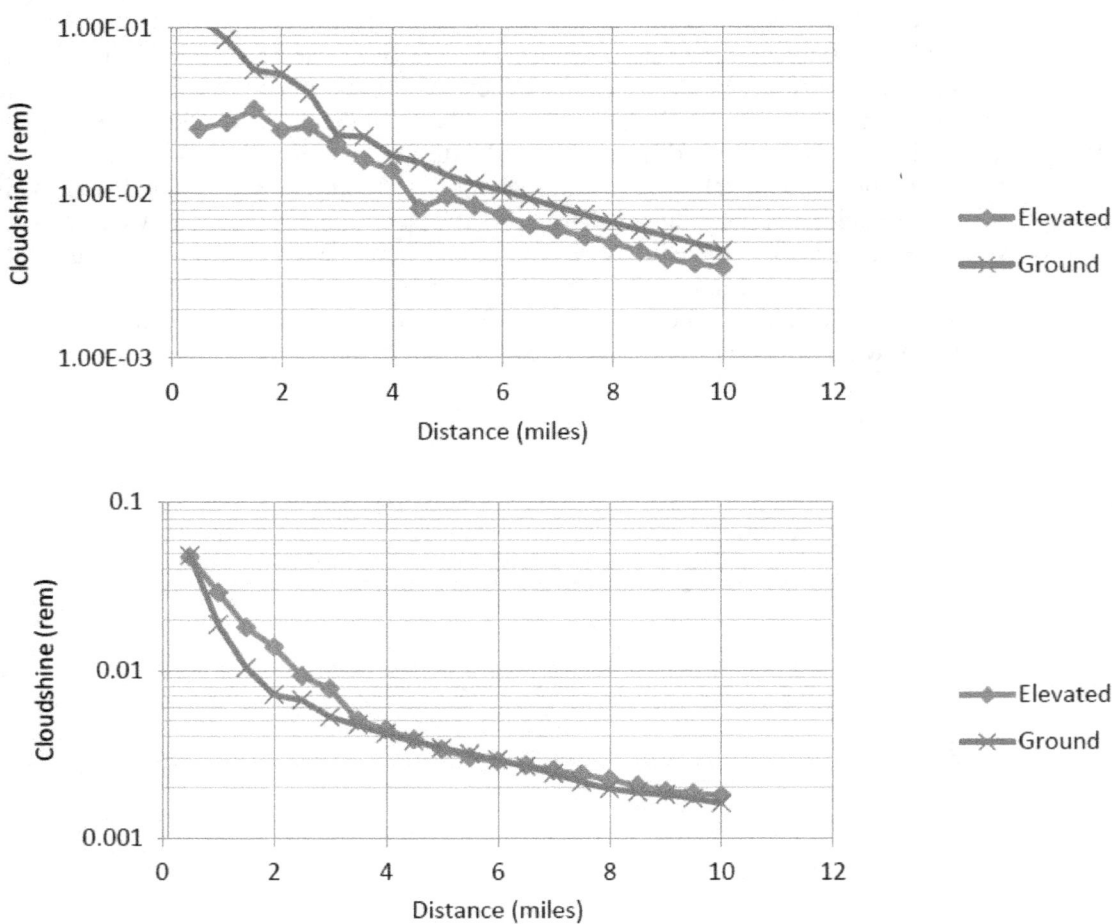

Figure 4-14 Numerical artifact in a cloudshine dose calculation associated with puff movement

4.11 Verification and Validation

Code verification and validation refer to a demonstration that a computer code represents the appropriate physical processes, that it is programmed correctly, and that it ultimately provides reasonable results compared to experimental data. RASCAL 4 is the start of the fourth generation of RASCAL codes and is a direct descendant of a series of computer codes that originated in the early 1970s. Many of the code modules associated with atmospheric transport and dispersion have been in existence since MESOI (Ramsdell et al., 1983). Dose calculations were introduced in MESORAD (Sherpelz, 1986; Ramsdell et al., 1988). Few changes in atmospheric transport and dispersion and dose calculations were made from MESORAD to RASCAL 2.0 (Athey et al., 1993). RASCAL 3.0 (Sjoreen et al., 2001) introduced an option to adjust wind fields for topography using a simple one-layer potential flow model. In RASCAL 4, the use of the potential flow model becomes standard practice. It is no longer optional for sites in the facility database. Wind field adjustments for topography cannot be made for sites that are not in the database. Section 6.5.2 discusses modifications to the wind field.

RASCAL 4 introduces significant upgrades to the atmospheric dispersion and deposition models and to the dose calculations. The revised dispersion and deposition algorithms were developed for use in

RATCHET, another descendant of MESORAD. RATCHET was developed as part of the HEDR project funded by the U.S. Centers for Disease Control and Prevention.

Several factors, including the following, contributed significantly to the selection of the RATCHET algorithms for use in RASCAL 4:

- An 18-member independent technical steering panel supervised the conduct of the HEDR project.

- Various organizations, including the National Institute of Statistical Sciences and the National Academy of Sciences, conducted project reviews.

- Appropriate specialists, including F.A Gifford, Jr., W.B. Petersen, W.F. Dabbert, W.B. Johnson, W.S. Lewellen, D.R. Randerson, R.P. Hosker, and S. R. Hanna, reviewed atmospheric transport, dispersion, and deposition algorithms.

- The HEDR project and code development followed strict quality assurance procedures.

- The HEDR project included an extensive model validation effort (Napier et al., 1994b).

- Subsequent dose reconstruction projects have used RATCHET and several of its concepts and components, e.g. Oak Ridge National Laboratory in Tennessee (Nair et al., 2000), Rocky Flats Plant in Colorado (Rood et al., 2002), Idaho National Engineering Laboratory in Idaho (Apostoaei, 2005a; Apostoaei et al., 2005b), and Mayak Reactor Plant in the Union of Soviet Socialist Republics (Glagolenko et al , 2008).

RASCAL and RATCHET were developed to address distinctly different problems. RASCAL addresses short-term accidental releases of radioactive material, whereas RATCHET addresses long-term routine releases of radioactive material (specifically I-131). As a result there are some differences in how the codes do their calculations. For example, RATCHET uses hourly meteorological data as input and generates daily output; RASCAL uses 15-minute meteorological data and generates output at 15-minute intervals. In addition, RATCHET includes stochastic components for treating uncertainty in parameters and input meteorological data; RASCAL does not include these components. However, no significant differences exist between the two codes in the treatment of atmospheric dispersion and deposition.

4.11.1 Verification

Verification of RASCAL 4 modules has been done in several ways. The two components of RASCAL 4 that do the atmospheric transport, dispersion, deposition, and dose calculations receive input from and generate files. The transfer of information from files has been checked numerous times by comparing data in the input files to data written to output files. Model parameters and control variables from input files are echoed in an output file. Hand calculations were used to verify elemental computational modules. The output file includes extensive lists of intermediate computational results. The intermediate results have been used to verify more complex calculations. In many cases, the plume model computational results have been used to verify the computational results of the puff model.

4.11.2 Validation

A model validation plan was developed for HEDR project and approved by the Technical Steering Panel (Napier, 1993). Validation of the atmospheric transport and deposition modules of RATCHET was done according to that plan (Napier et al., 1994a). Most of the validation tests involved several of the project component models (e.g., the source term model, the atmospheric models, and vegetation model). Two of the validation tests started with reasonably well-known source terms and are particularly useful in demonstrating the validity of the RATCHET atmospheric model components that are included in RASCAL.

The first of these tests was modeling an unplanned release of about 72 curies of I-131 in September 1963. Contemporary environmental measurements (Soldat, 1965) included measurements of I-131 in grass and milk at two farms about 20 miles south of the release point. Figures 4-15 and 4-16 compare the HEDR model predictions of I-131 in grass and milk with the measured values. The HEDR models were stochastic, and 100 sets of model predictions were produced. These figures show the range of model predictions and the median prediction. This range shows the uncertainty in model estimates. This uncertainty reflects the combined uncertainties in wind direction, wind speed, stability, interception of deposited iodine by grass, and uptake and accumulation of iodine in the cows. The last two areas of uncertainty were added by HEDR models that used RATCHET output; they were not included in RATCHET . Note that variability of the measured data decreases when a time and spatially integrating sampler (cow) is used. Two children, a boy and a girl, who were living on Farm B drank milk from the cows. The median HEDR model estimates of thyroid doses were 45.3 millirad (mrad) for the boy and 11.3 mrad for the girl. Contemporary estimates of thyroid dose burden were 35 mrad for the boy and 9 mrad for the girl. Although none of these results directly validates the atmospheric transport and dispersion algorithms in RATCHET, they do validate the algorithms indirectly because RATCHET was the initial model in the sequence of models that produced the results. If the RATCHET algorithms contained significant errors the results shown would reflect those errors.

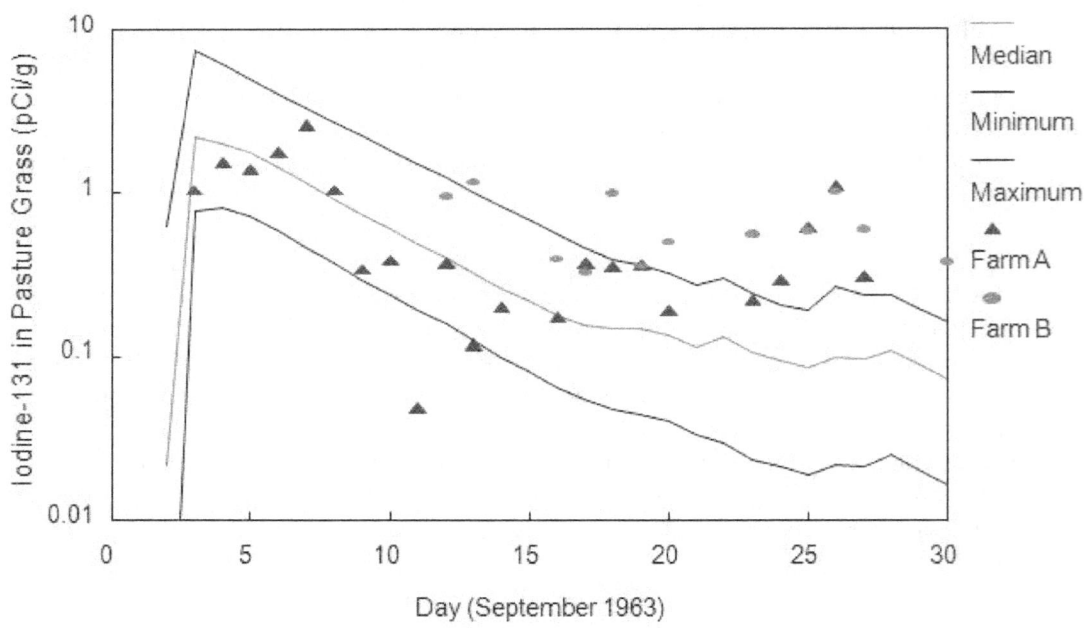

Figure 4-15   Comparison of measured I-131 in grass to HEDR model predictions

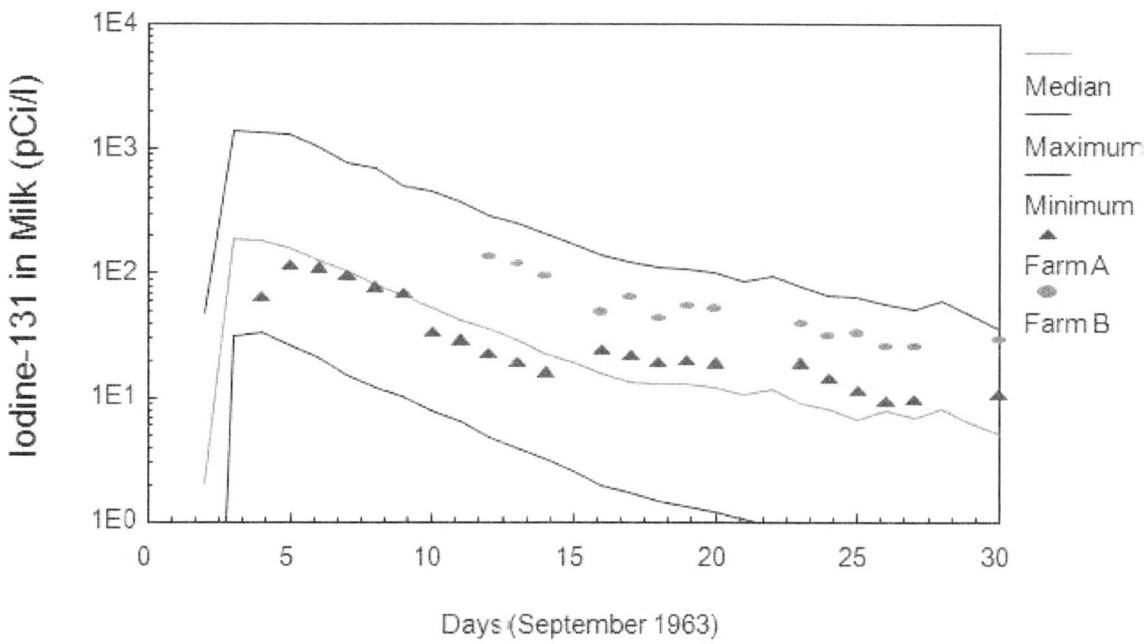

Figure 4-16   Comparison of measured I-131 in milk to HEDR model predictions

The second test of RATCHET involved modeling dispersion of krypton-85 (Kr-85) from processing reactor fuel at the Plutonium and Uranium Recovery by Extraction (PUREX) Plant at Hanford Site from late 1983–1987. The Kr-85 was released from a 200-foot stack. Release rates were estimated from the log books for the PUREX Plant. The PUREX Plant was assumed to be the only source of Kr-85. However, Hanford Site annual environmental reports indicate that there were Kr-85 releases from other sources in 1985 and 1986. A cryogenic monitoring network was in place at Hanford during this period. In 1984, measurements were made at four locations. By 1987, the network had expanded to include measurements at 11 locations. However, complete data are not available for all locations. The measurements consisted of samples accumulated over periods ranging from 14 to 38 days with 28-day samples being typical. Meteorological data for the validation tests consisted of a complete meteorological observations for the Hanford Meteorological Station and hourly wind speed and direction data for an additional 24 locations on and near the Hanford Site.

Figures 4-17–4-19 present comparisons between RATCHET predictions of Kr-85 concentrations and monitoring data. Figure 4-17 compares long-term average concentrations for the 11 monitoring locations. Most of the locations are in the southeast quadrant from the release point at distances between 18 and 35 miles. Yakima, WA, is about 50 miles west of the release point and is separated by Rattlesnake Ridge from the Hanford Site. Othello, WA, is about 24 miles northeast of the Hanford Site. The two 300 Trench monitoring locations are collocated. Those two locations and the Fir Road location map to the same receptor node in the RATCHET code output. A 24-picocurie (pCi) background concentration has been added to the RATCHET predictions.

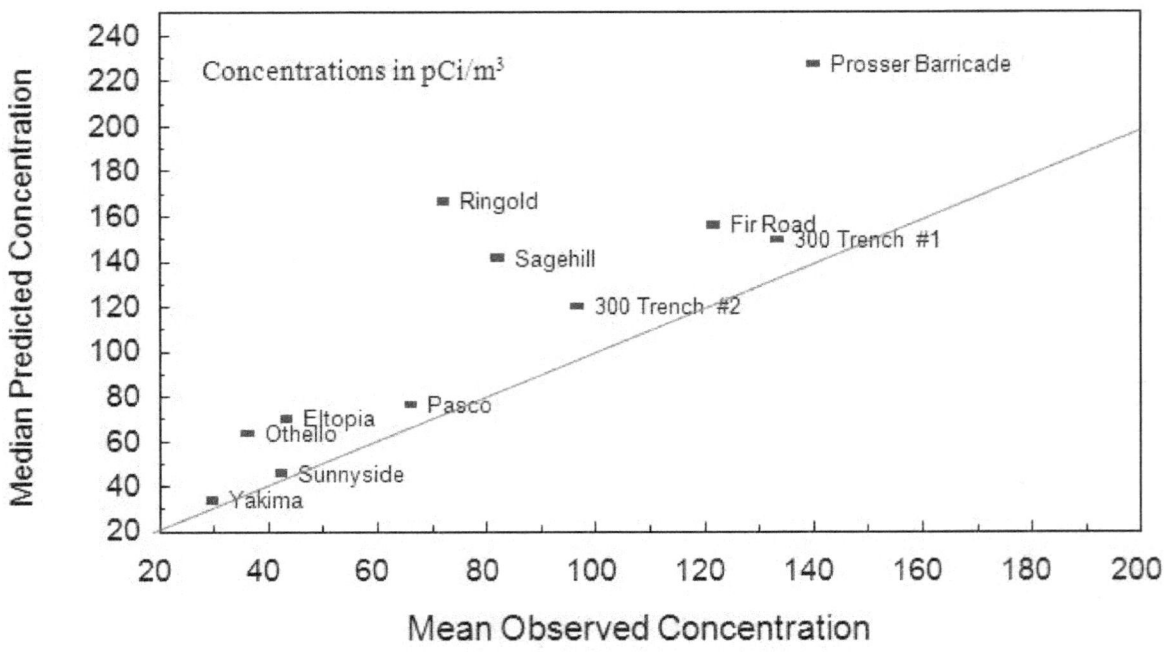

Figure 4-17 Modeled and measured 4-year average Kr-85 air concentrations at 11 sampling locations on and near the Hanford Site

Figure 4-18 compares RATCHET estimates of average Kr-85 air concentrations at a receptor location about 18 miles southeast of the release point to data from three monitoring locations near the receptor location. The range of RATCHET shown in Figure 4-18 shows the uncertainty associated with transport and dispersion. The figure does not include uncertainty in either the quantities of Kr-85 released or the timing of the releases. In addition, the figure does not show uncertainty in the measured values. Uncertainty in the Kr-85 activity because of counting has been reported as ±10 percent. Larger uncertainties appear to be associated with the sampling itself. For example, note the large difference in the concentrations reported for the collocated 300 Trench monitors at about day 1,225 and again at about day 1,275.

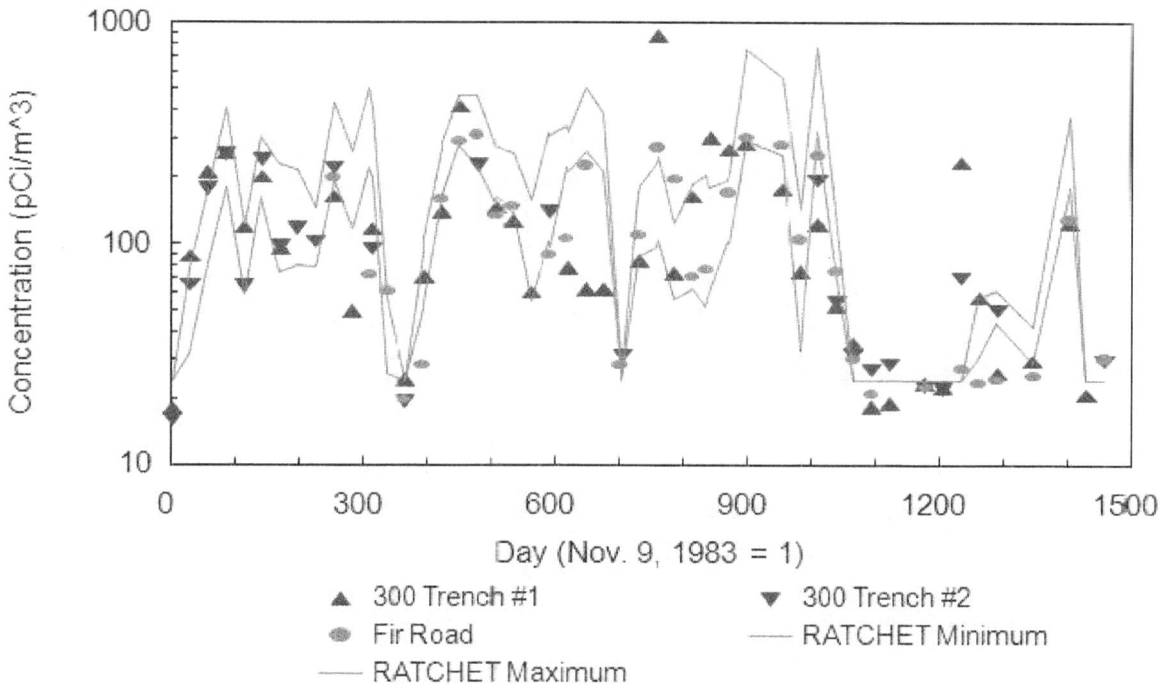

Figure 4-18 Comparison of observed nominal monthly average observed Kr-85 concentrations to the range of concentrations estimated by RATCHET

Figure 4-19 illustrates the variation in concentration estimates among Kr-85 samples. The data shown in the figure are for nominal 30-day samples collected during common periods by the 300 Trench monitors. Differences on the order of a factor of 2 are not uncommon.

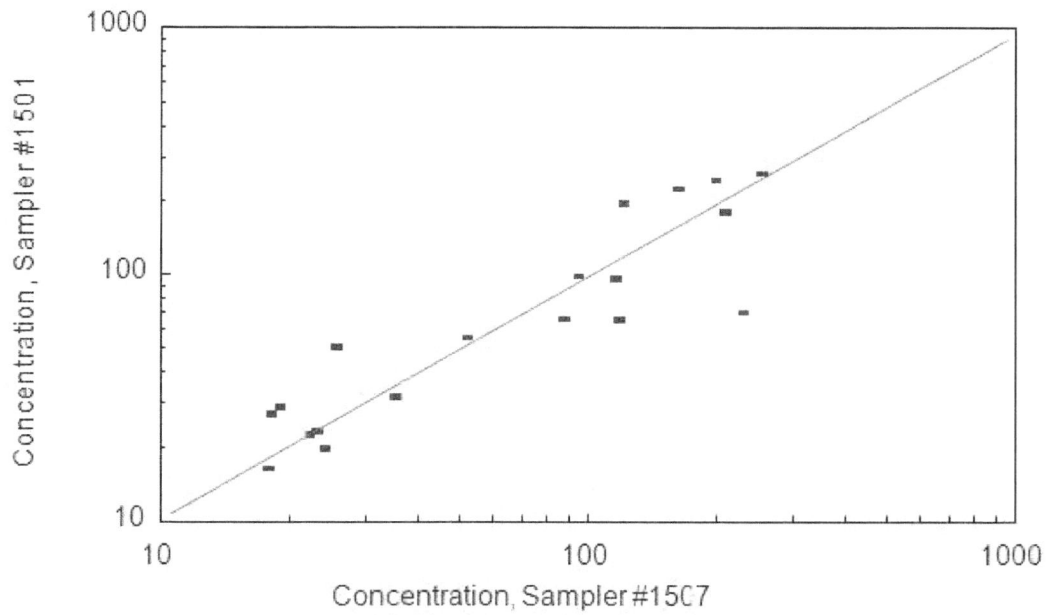

Figure 4-19 Variation in Kr-85 monitor concentration estimates

Finally, Figure 4-20 presents the cumulative distribution of the ratios of the median monthly average Kr-85 concentrations predicted by RATCHET to the observed values for the Kr-85 data set. The "Full Met. Data" (full meteorological data) curve is for model runs made using all available meteorological data; the "Limited Met. Data" (limited meteorological data) curve is for model runs made using a subset of the available meteorological data set that approximated the meteorological data available for evaluating I-131 releases at the Hanford Site in the 1940s. The figure shows that about 75 percent of the observed concentrations were within a factor of 2 of the median concentration estimated by RATCHET.

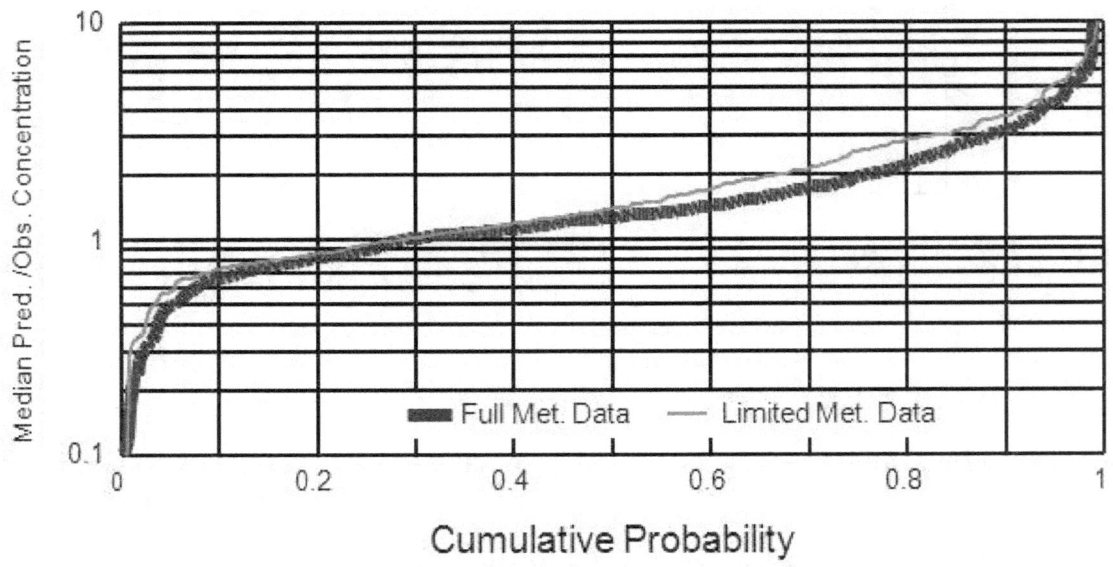

Figure 4-20 Cumulative frequency distribution of predicted to observed concentration ratios for RATCHET for the Kr-85 data set

In preparation for Phase II of the Historical Public Exposure Studies at the Rocky Flats Plant in Colorado, Rood et al. (1999) evaluated five atmospheric models, including RATCHET. Rood et al. (1999) performed this evaluation using data from 12 tracer release experiments at the Rocky Flats Plant in 1991. Among the measures used to evaluate the models were comparisons between predicted and observed concentrations paired in space at distances of 8 and 16 km from the release point. RATCHET had a higher correlation between predicted and observed concentrations and a higher percentage of predicted concentrations within a factor of 5 of the observed value at both 8 and 16 km than the other 4 models did.

To this point, the validation discussion has compared the results of model calculations to the results of field measurements. Another method of validating a model is to compare the results of a relatively simple model to the results of other models, especially if the model used as the standard contains more detailed representations of the important physical processes and offers higher resolution results.

The NRC sponsored such a comparison in 2003 and 2004 (Molenkamp et al., 2004). The predictions of atmospheric dispersion portions of the MACCS2 (Chanin and Young., 1998) and RASCAL 3.0 (Sjoreen et al., 2001) computer codes were compared predictions of the Lawrence Livermore National Laboratory's LODI/ADAPT model (Nasstrom et al., 2000; Sugiyama and Chan, 1998) for 610 hypothetical 30-minute releases of a nondepositing and a depositing tracer. Predictions of the RATCHET model were also compared to the LODI/ADAPT predictions in anticipation of including the RATCHET dispersion and deposition algorithms in RASCAL. In terms of complexity, the atmospheric models increase in complexity from MACCS2 (transport and dispersion based on release-point meteorology) to

RASCAL and RATCHET (two-dimensional temporally and spatially varying meteorology) to LODI/ADAPT (three-dimensional temporally and spatially varying meteorology).

The site selected for the comparison was the U.S. Department of Energy's Atmospheric Radiation Measurement Program Southern Great Plains site in central Oklahoma and Kansas. This site was selected primarily because of the extensive meteorological data set available. The comparison assumed that all releases occurred at a height of 50 meters with a 1-megawatt heat flux. Comparison metrics were the exposure and deposition averaged over 360 degrees around the release point at nominal distances of 9.5, 19.5, and 49.5 miles and exposure and deposition in each of 16 sectors at the same nominal distances. In general, the predictions of the simpler models for each of these metrics were within a factor of 2 of the LODI/ADAPT predictions. RASCAL predictions tended to be higher than the LODI/ADAPT predictions, and RATCHET predictions tended to be lower.

Other metrics related to this study have significance related to emergency consequence assessment. These metrics are related to the time required to prepare the meteorological data and to perform the computations. MACCS2, RASCAL, and RATCHET were run on personal computers. ADAPT and LODI were run on a Digital Equipment Corporation computer. Meteorological data preparation for the MACCS2 code involved performing routine quality assurance checks, filling in missing data, and formatting the data for use by the code. The initial steps in meteorological data preparation for RASCAL and RATCHET were the same. However, RASCAL and RATCHET use spatial data; therefore, they required an additional data preparation step—preparation of the meteorological data fields. Preparation of the hourly meteorological data fields for the entire year took approximately 16 minutes of computer time. Finally, preparation of the meteorological data files for ADAPT took several weeks of data review and manipulation of the raw meteorological data. It then took ADAPT about 98 hours to prepare the meteorological data input for LODI, which did the transport and dispersion calculations. The times spent doing the actual transport and dispersion calculations was as follows: (1) MACCS2 took less than a minute, (2) RASCAL took 46 minutes, (3) RATCHET took 65 minutes, and (4) LODI took 1,403 minutes. Table 4-4 summaries these times.

Table 4-4   Model Computational Times

MODEL	METEOROLOGICAL PREPROCESSING	TRANSPORT AND DISPERSION
MACCS2	---	<1 minutes[a]
RASCAL	16 minutes[b]	46 minutes[c]
RATCHET	16 minutes[b]	65 minutes[c]
ADAPT	5,852 minutes[b]	---
LODI	---	1,403 minutes[a]
(a) 610 releases.		
(b) Hourly wind fields for 1 year.		
(c) Total time for the three runs of 610 releases (one for each grid).		

4.12 References

Aoyama, M., K. Hirose, Y. Suzuki, H. Inoue, and Y. Sugimura. 1986. "High-Level Radioactive Nuclides in Japan in May," *Nature,* 321(6073):819–820.

Apostoaei, A.I. 2005a. "Testing Prediction Capabilities of an [131]I Terrestrial Transport Model by Using Measurements Collected at the Hanford Nuclear Facility," *Health Physics,* 88(5):439–58.

Apostoaei, A.I., B.A. Thomas, D.C. Kocher, and F.O. Hoffman. 2005b, "Doses to the Public from Atmospheric Releases of Radionuclides from the Idaho Chemical Processing Plant at the Idaho National Engineering Laboratory (1957–1959)," Report to the Centers for Disease Control and Prevention, Contract No. 200-2002-00367, SENES Oak Ridge, Inc., Oak Ridge, TN.

Athey, G.F. A.L. Sjoreen, J.V. Ramsdell, and T. J. McKenna. 1993. "RASCAL Version 2.0 User's Guide," Volume 1, Revision 1, NUREG/CR-5247. U.S. Nuclear Regulatory Commission, Washington, DC.

Bander, T.J. 1982. "PAVAN: An Atmospheric Dispersion Program for Evaluating Design-Basis Accidental Releases for Radioactive Materials from Nuclear Power Stations," NUREG/CR-2858, U.S. Nuclear Regulatory Commission, Washington, DC, November.

BIOMOVS. 1990. (Biospheric Model Validation Study), "BIOMOVS Technical Report Scenario A4, Appendix 2," (Input data and documentation of observations), Stockholm, Sweden.

Bondietti, E.A., and J.N. Brantley. 1986. "Characteristics of Chernobyl Radioactivity in Tennessee," *Nature,* 322(6077):313–314.

Cambray, R.S., P.A. Cawse, J.A. Garland, J.A.B. Gibson, P. Johnson, G.N.J. Lewis, D. Newton, L. Salmon, and B.O. Wade. 1987. "Observations on Radioactivity from the Chernobyl Accident," *Nuclear Energy*, 26(2):77–101.

Cember, H. 1996. *Introduction to Health Physics*, Third Edition, McGraw-Hill, New York, NY.

Chanin, D.I., and M.L. Young. 1998. "Code Manual for MACCS2: Volume 1, 'User's Guide,'" NUREG/CR-6613, U.S. Nuclear Regulatory Commission, Washington, DC.

Draxler, R.R. 1984. "Diffusion and Transport Experiments," *Atmospheric Science and Power Production,* Ed., D. Randerson, DOE/TIC-27601, U.S. Department of Energy, Washington, DC.

Eckerman, K.F., A.B. Wolbarst, and A.B.C. Richardson. 1988. "Limiting Values of Radionuclide Intake and Air Concentration and Dose Conversion Factors for Inhalation, Submersion, and Ingestion," Federal Guidance Report No. 11, EPA-520/1-88-020, U.S. Environmental Protection Agency, Washington, DC.

Eckerman, K.F., and J.C. Ryman. 1993. "External Exposure to Radionuclides in Air, Water, and Soil," Federal Guidance Report No. 12, EPA-402-R-93-081, U.S. Environmental Protection Agency, Washington, DC.

Eckerman, K.F., R.W. Leggett, C.B. Nelson, J.S. Puskin, and A.C.B. Richardson. 1999. "Cancer Risk Coefficients for Environmental Exposure to Radionuclides," Federal Guidance Report No. 13, EPA-402-R-99-001, U.S. Environmental Protection Agency, Washington, DC.

Eckerman, K.F., and R.W. Leggett. 2002. "Dose and Risk Coefficient File Package for FGR-13," FGR13PAK, Oak Ridge National Laboratory, Oak Ridge, TN.

General Public Utilities Service Corporation (GPUSC). 1972. "Atmospheric Diffusion Experiments with SF6 Tracer Gas at Three Mile Island Nuclear Station under Low Wind Speed Inversion Conditions," Amendment No. 24, Docket No. 50-289, General Public Utilities Service Corporation, Parsippany, NJ.

Glagolenko, Yu.V., E.G. Drozhko, Yu.G. Mokrov, S.I. Rovny, D.A. Beregich, P.M. Stukalov, I.A. Ivanov, A.I. Alexakhin, L.R. Anspaugh, and B.A. Napier. 2008. "Reconstruction of External Doses to Ozyorsk Residents Due to Atmospheric Releases of Inert Radioactive Gases from the Stacks of the 'Mayak PA' Reactor Production from 1948 to 1989," *Radiation Safety Problems* (*Mayak Production Association Scientific Journal Special Issue*), 20–31.

Gifford, F. A. 1976. "Turbulent Diffusion-Typing Schemes: A Review." *Nuclear Safety,* 17 (1): 68–86.

Gifford, F.A. 1982. "Horizontal Diffusion in the Atmosphere: A Lagrangian-Dynamical Theory," *Atmospheric Environment* 16(3):505–512.

Hanna, S.R., G.A. Briggs, and R.P. Hosker. 1982. "Handbook on Atmospheric Diffusion," DOE/TIC-11223, U.S. Department of Energy, Washington, DC.

Healy, J.W., and R.E. Baker. 1968. "Radioactive Cloud-Dose Calculations." *Meteorology and Atomic Energy 1968.* Ed. D. Slade, TID-24190, U.S. Atomic Energy Agency, Washington, DC.

Healy, J. W. 1984. "Radioactive Cloud-Dose Calculations." *Atmospheric Science and Power Production,* Ed., D. Randerson. DOE/TIC-27601, U.S. Department of Energy, Washington, DC.

International Atomic Energy Agency (IAEA). 2006. *Dangerous Quantities of Radioactive Materials, EPR-D-Values,* Vienna, Austria.

International Commission on Radiological Protection (ICRP). 1991. "1990 Recommendations of the International Commission on Radiological Protection," ICRP Publication 60, *Annals of the ICRP,* 21:1-3.

Irwin, J.S. 1983. "Estimating Plume Dispersion—A Comparison of Several Sigma Schemes," *Journal of Climate and Applied Meteorology,* 22(1):92–114.

Ludwick, J.D. 1964. "Investigation of the Nature of I-131 in the Atmosphere. In *Hanford Radiological Sciences Research and Development Annual Report for 1963*, HW-81746, pp. 3.58–3.68, Eds., C.C. Gamertsfelder and J.K. Green. General Electric Company, Hanford Atomic Products Operation, Richland, WA.

Ludwick, J.D. 1967. ., "A Portable Boom-Type Air Sampler." In *Pacific Northwest National Laboratory Annual Report for 1966 to the USAEC Division of Biology and Medicine, Volume II: Physical Sciences, Part 1,* BNWL-481, pp. 87–92, *Atmospheric Sciences,* Eds., D.W. Pearce and M.R. Compton, Pacific Northwest National Laboratory, Richland, WA.

Molenkamp, C.R., N.E. Bixler, C.W. Morrow, J.V. Ramsdell, Jr., and J.A. Mitchell. 2004. "Comparison of Average Transport and Dispersion Among a Gaussian, a Two-Dimensional, and a Three-Dimensional Model," NUREG/CR-6853. U.S. Nuclear Regulatory Commission, Washington, DC.

Mueck, K. 1988. "Variations in Activity Concentration and Radionuclide Ratio in Air after the Chernobyl Accident and Its Relevance to Inhalation Dose Estimate, *Radiation Protection Dosimetry,* 22(4):219-229.

Nair, S.K., A.I. Apostoaei, and F.O. Hoffman. 2000. "A Radioiodine Speciation, Deposition, and Dispersion Model with Uncertainty Propagation for the Oak Ridge Dose Reconstruction," *Health Physics*, 78(4):394–413.

Napier, B.A., R.O. Gilbert, J.C. Simpson, J.V. Ramsdell, Jr., M.E. Thiede, and W.H. Walters. 1993. "HEDR Model Validation Plan," PNWD-2156 HEDR, Pacific Northwest National Laboratory, Richland, WA.

Napier, B.A., J.C. Simpson, P.W. Eslinger, J.V. Ramsdell, Jr., M.E. Thiede, and W.H. Walters. 1994. "Validation of HEDR Models," PNWD-2221 HEDR, Pacific Northwest National Laboratory, Richland, WA.

Nasstrom, J.S., G. Sugiyama, J.M. Leone, Jr., and D.L. Ermak, 2000. "A Real-Time Atmospheric Dispersion Modeling System," *Preprint of the Eleventh Joint Conference on the Applications of Air Pollution Meteorology*, Long Beach, CA, January 9–14, 2000, American Meteorological Society, Boston, MA. (Available URL: http://www.llnl.gov/tid/lof/documents/ pdf/237149.pdf)

Panofsky, H.A., and J.A. Dutton. 1984. *Atmospheric Turbulence*, J. Wiley & Sons, New York, NY.

Panofsky, H.A., H. Tennekes, D.H. Lenschow, and J.C. Wyngaard. 1977. "The Characteristics of Turbulent Velocity Components in the Surface Layer under Convective Conditions," *Boundary Layer Meteorology* 11(3):355–361.

Perkins, R.W. 1963. "Studies of Radioiodine and Other Fallout Radionuclides in Air," *Hanford Radiological Sciences Research and Development Annual Report for 1962*, HW-77609, pp. 3.36–3.48. Richland, WA.

Perkins, R.W. 1964. "Physical and chemical forms of I-131 from fallout and chemical processing plants," *Hanford Radiological Sciences Research and Development Annual Report for 1963*, HW-81746 pp. 3.55-3.58, Eds., C.C. Gamertsfelder and J.K. Green. General Electric Company, Hanford Atomic Products Operation, Richland, WA.

Petersen, W.B., and L.G. Lavdas. 1986. "INPUFF 2.0—A Multiple Source Gaussian Puff Dispersion Algorithm: User's Guide," EPA-600/8-86/024, Atmospheric Sciences Research Laboratory, U.S. Environmental Protection Agency, Washington, DC.

Press, W.H., B.P. Flannery, S.A. Teukolsky, and W.T. Vetterling. 1986. *Numerical Recipes: The Art of Scientific Computing,* Cambridge University Press, Cambridge, United Kingdom.

Ramsdell, Jr., J.V., G.F. Athey, and C.S. Glantz. 1983. "MESOI Version 2.0: An Interactive Mesoscale Lagrangian Puff Dispersion Model with Deposition and Decay," NUREG/CR-3344, U.S. Nuclear Regulatory Commission, Washington, DC.

Ramsdell, Jr., J. V., et al. 1988. 'The MESORAD Dose Assessment Model, Volume 2: Computer Code." Vol. 2., NUREG/CR-4000, U.S. Nuclear Regulatory Commission, Washington, DC

Ramsdell, Jr., J. V. 1990. "Diffusion in Building Wakes for Ground-Level Releases." *Atmospheric Environment,* 24B:377–88.

Ramsdell, Jr, J. V., C. A. Simonen, and K. W. Burk. 1994. "Regional Atmospheric Transport Code for Hanford Emission Tracking (RATCHET)." PNWD-2224 HEDR, Battelle, Pacific Northwest Laboratories, Richland, Wash.

Ramsdell, Jr., J. V., C. A. Simonen, K. W. Burk, and S. A. Stage. 1996. "Atmospheric Dispersion and Deposition of 131I Released From The Hanford Site." *Health Physics* 71:568-577.

Ramsdell, Jr., J. V., and C. J. Fosmire. 1998. "Estimating Concentrations in Plumes Released in the Vicinity of Buildings: Model Development." *Atmospheric Environment,* 32:1663–17.

Randerson, D. 1984. *Atmospheric Science and Power Production.* DOE/TIC-27601. U.S. Department of Energy, Washington, DC.

Rood, A. S. G. G. Killough, and J.E. Till. 1999. Evaluation of Atmospheric Transport Models for Use in Phase II of the Historical Public Exposures Studies at the Rocky Flats Plant. *Risk Analysis*, 19(4): 559-576.

Rood, A.S., H.A. Grogan, and J.E. Till. 2002. "A Model for Comprehensive Assessment of Exposure and Lifetime Cancer Incidence Risk from Plutonium Released from the Rocky Flats Plant, 1953–1989," *Health Physics,* 82(2):182–212.

Rood, A.S., G.G. Killough, and J.E. Till, "Evaluation of Atmospheric Transport Models for Use in Phase II of the Historical Public Exposures Studies at the Rocky Flats Plant," *Risk Analysis*, 19(4):559–576, 1999.

Sagendorf, J.F., J.T. Goll, and W.F. Sandusky. 1982. "XOQDOQ: Computer Program for the Meteorological Evaluation of Routine Effluent Releases at Nuclear Power Stations," NUREG/CR-4330, U.S. Nuclear Regulatory Commission, Washington, DC.

Sandia National Laboratories (SNL). 2009, "TurboFRMAC 2009," Sandia National Laboratories, Albuquerque, NM.

Sandia National Laboratories (SNL). 2010. "FRMAC Assessment Manual: Volume 1, 'Overview and Methods,'" SAND2011-1405P, Sandia National Laboratories, Albuquerque, NM.

Scherpelz, R.I., T.J. Bander, G.F. Athey, and J.V. Ramsdell. 1986. "The MESORAD Dose Assessment Model: Volume 1, 'Technical Basis,'" NUREG/CR-4000, U.S. Nuclear Regulatory Commission, Washington, DC.

Sehmel, G.A. 1984. "Deposition and Resuspension," *Atmospheric Science and Power Production,* Ed., D. Randerson, DOE/TIC-27601, U.S. Department of Energy, Washington, DC.

Seinfeld, J.H. 1986. *Atmospheric Chemistry and Physics of Air Pollution,* John Wiley & Sons, New York, NY.

Shipler, D.B., B.A. Napier, W.T. Farris, and M.D. Freshley. 1996. "Hanford Environmental Dose Reconstruction Project—An Overview," *Health Physics,* 71:532–544.

Sjoreen, A.L., J.V. Ramsdell, Jr., T.J. McKenna, S.A. McGuire, C. Fosmire, and G.F. Athey. 2001. "RASCAL 3.0: Description of Models and Methods," NUREG-1741, U.S. Nuclear Regulatory Commission, Washington, DC.

Slade, D.H. 1968. "Meteorology and Atomic Energy 1968," TID-24190, U.S. Atomic Energy Agency, Washington, DC.

Slinn, W.G.N. 1984. "Precipitation Scavenging," *Atmospheric Science and Power Production.* Ed., D. Randerson, DOE/TIC-27601, U.S. Department of Energy, Washington, DC.

Soffer, L., S.B. Burson, C.M. Ferrell, R.Y. Lee, and J.N. Ridgely. 1995. "Accident Source Terms for Light-Water Nuclear Power Plants," NUREG-1465, U.S. Nuclear Regulatory Commission, Washington, DC.

Soldat, J.K. 1965. "Environmental Evaluation of an Acute Release of I^{131} to the Atmosphere," *Health Physics,* 11(10):1009–1015.

Start, G.E., J.F. Cate, C.R. Dickson, N.R. Ricks, G.R. Ackerman, and J.F. Sagendorf. 1978. "Rancho Seco Building Wake Effects on Atmospheric Diffusion," NUREG/CR-0456, U.S. Nuclear Regulatory Commission, Washington, DC.

Sugiyama, G., and S.T. Chan, 1998. "A New Meteorological Data Assimilation Model for Real-Time Emergency Response," *Preprint of the Tenth Joint Conference on the Applications of Air Pollution Meteorology*, Phoenix, AZ, January 11–16, 1998, American Meteorological Society, Boston, MA, pp. 285–289. ((Available URL: http://www.llnl.gov/tid/lof/documents/pdf/232515.pdf)

Thiessen, K.M., B.A. Napier, B. Filistovic, T. Homma, B. Kanyar, P. Krajewski, A.I. Kryshev, T. Nedveckaite, A. Nenyei, T.G. Sazykina, U. Tveten, K.L. Sjoblom, and C. Robinson. 2005. "Model Testing Using Data on 131I Released from Hanford," *Journal of Environmental Radioactivity* 84:211–224.

Thuillier, RH. and R.M. Mancuso. 1980. *Building effects on effluent dispersion from roof vents at nuclear power plants.* EPRI NP-1380, Electric Power Research Institute, Palo Alto, CA.

Thuillier, R.H. 1988. "Diffusion in Building Wakes," *Proceedings of the American Nuclear Society Topical Meeting on Emergency Response—Planning, Technologies, and Implementation*, Charleston, SC., CONF-880913, U.S. Department of Energy, Washington,

U.S. Environmental Protection Agency (EPA). 1992. "Manual of Protective Action Guides and Protective Actions for Nuclear Incidents," EPA-400-R-92-001, U.S. Environmental Protection Agency, Washington, DC.

Voilleque, P.G. 1979. "Iodine Species in Reactor Effluents and in the Environment," EPRI-NP-1269, Electric Power Research Institute, Palo Alto, CA.

5. UF$_6$ TRANSPORT AND DIFFUSION MODEL

RASCAL 4 contains a special version of the plume model (Chapter 4) that has been modified to treat releases of uranium hexafluoride (UF$_6$). The modifications include the introduction of a dense gas model to treat the gravitationally driven spread of UF$_6$ releases, a chemical/thermodynamic model to treat the reaction of UF$_6$ with water (both liquid and vapor) in the atmosphere, and a plume rise model to treat the vertical displacement of hydrogen fluoride/uranyl fluoride (HF/UO$_2$F$_2$) plumes when their densities become less than the density of air. The chemical reaction of UF6 and water is:

$$UF_6 - 2H_2O \rightarrow UO_2F_2 + 4HF + heat \tag{5-1}$$

The implementation of dense gas and chemical/thermodynamic models is done in two control volumes, one for UF$_6$ and a second for HF and UO$_2$F$_2$. "Control volumes," as used in thermodynamics, are volumes in which mass, energy, moisture, and other parameters are evaluated taking into account the quantities moving into and out of the volume. The control volumes move downwind at the speed of the wind 1 meter above ground level. The size of the control volumes are initially defined by the release rates of UF$_6$, HF, and UO$_2$F$_2$. As the control volumes move downwind, gravitational settling deforms the volume of UF$_6$, and air and water vapor are mixed into the UF$_6$ volume. The models assume that chemical reaction occurs instantaneously as the mixing takes place. The reaction results in a decrease of mass and volume of the control volume containing UF$_6$ and an increase in the mass and volume in the HF/UO$_2$F$_2$ control volume. The temperatures of these two control volumes are assumed to be identical and are determined from the initial temperature of the released material, the air temperature, and the heat of reaction of UF$_6$ and water.

The output of the dense gas and chemical/thermodynamic model calculations is used as input to atmospheric dispersion and deposition calculations. This input is a function of the distance from the release point to the point at which all the UF$_6$ has been converted to HF and UO$_2$F$_2$. After the UF$_6$ is gone, the HF and UO$_2$F$_2$ source terms continue to decrease with distance to account for deposition as described in Chapter 4.

5.1 UF$_6$ Model Assumptions and Equations

The following assumptions were made in the development of the UF$_6$ model:

- The UF$_6$ plume is released at or near ground level. (Elevated releases are not modeled.)

- The UF$_6$ release rate and density define an initial UF$_6$ control volume.

- The initial cross-section of the UF$_6$ control volume is square with the cross-sectional are given by:

$$A_{UF6} = \frac{Q'_{UF6}}{\rho_{UF6}u}, \tag{5-2}$$

where:

> · A_{UF6} = cross-sectional area (m^2)
> Q'_{UF6} = UF$_6$ release rate (g/s),
> ρ_{UF6} = UF$_6$ density (g/m^3),
> u = wind speed at a height of 1 meter (m/s),

If the release includes HF and UO_2F_2 in addition to UF_6, the cross-sectional area of the initial control volume is:

$$A_{cv} = \frac{V'_{UF6} + V'_{HF} + V'_{air}}{u},$$ (5-3)

where:

V'_{UF6} = the release rate of UF_6 (m³/s)
V'_{HF} = the release rate of HF (m³/s)
V'_{air} = the volume flow of air that would be needed to generate the HF flow from a reaction of the air with UF_6 (m³/s)
u = wind speed at a height of 1 meter (m/s)

- The UF_6 plume does not diffuse.

- Gravitational slumping of the UF_6 determines the deformation of the UF_6 control volume.

- The rate of change of the UF_6 control volume width is:

$$\frac{dw_{UF6}}{dt} = k\left[g\frac{(\rho_{UF6} - \rho_{air})}{\rho_{UF6}}H_{UF6}\right],$$ (5-4)

where:

w_{UF6} = UF_6 control volume width (m)
t = time (s)
k = a slumping constant (dimensionless)
g = gravitational constant (m/s²)
ρ_{air} = density of air (g/m³)
H_{UF6} = thickness of the control volume (m)

- The slumping constant has a theoretical value of 1.4 ($2^{1/2}$) (Eidsvik, 1980), but it may be given a lower value to account for surface resistance or to tune the model. The current version of the UF_6 model in RASCAL uses a value of 1.3 as the default.

- Air is entrained into the UF_6 control volume only through the top. Entrainment through the sides is negligible because, after only a few seconds, the area of the top of the volume is much larger than the area of the sides.

- The rate of entrainment of air into the UF_6 is:

$$\frac{dV_{air}}{dt} = u_e w_{UF6} u,$$ (5-5)

where:

V_{air} = air entrainment rate (m³/s)
u_e = an entrainment velocity (m/s)

- The entrainment velocity u_e is:

$$u_e = \frac{\rho_{air} u_*^3}{(\rho_{UF6} - \rho_{air}) g h_{uf6}},$$ (5-6)

where u_* is a scaling velocity (m/s) associated with atmospheric turbulence.

- A combination of the water vapor in the entrained air and precipitation entering the UF_6 control volume determines the water available for reaction with UF_6.

- The water available for reaction is:

$$m_{H2O} = \rho_{H2Ov} V_{air} + p_r w_{UF6} u \Delta t \rho_{H2Ol},$$ (5-7)

where:

Δt = the duration of the time step (s)
m_{H2O} = the rate at which water (H_2O) becomes available for reaction (g/s)
ρ_{H2Ov} = density of water vapor in the ambient air (g/m^3)
p_r = precipitation rate (m/s)
ρ_{H2Ol} = density of liquid water (g/m^3)

- The reaction between UF_6 and water occurs at the top of the UF_6 control volume. The volume of UF_6 involved in the reaction is subtracted from the UF_6 control volume, and the masses of air, HF, and UO_2F_2 are added to the HF/UO_2F_2 control volume. The volumes of air and HF increase the volume of the HF/UO_2F_2 control volume. The UO_2F_2 formed in the UF_6/H_2O reaction is present as small particles that have negligible volume. The temperatures and volumes of the control volumes are adjusted to conserve enthalpy in a constant pressure reaction.

- Potential heat exchange with the ground and the possible reaction of UF_6 with water on the ground surface are negligible.

- The ground is a sink for UF_6 that may be deposited on the ground. Any UF_6 condensing in the UF_6 control volume deposits on the ground. In addition, 25 percent of the UO_2F_2 formed in the UF_6/H_2O reaction deposits at the time of the reaction, unless the UF_6 is released in a fire. Wet deposition of UF_6 is not modeled because of the assumption that all water entering the UF_6 control volume reacts with UF_6 to produce HF and UO_2F_2.

- If UF_6 is released within a building, it reacts with water vapor within the building, and the release to the environment consists of only HF and UO_2F_2. In this instance, RASCAL does not do thermodynamic calculations, and the transport and dispersion calculations do not include plume rise.

5.2 Chemical/Thermodynamic Model

The chemical/thermodynamic model in the UF_6 plume model is based on the description contained in NUREG/CR-4360, "Calculational Methods for Analysis of Postulated UF_6 Releases," (Williams, 1985). The initial release to the atmosphere may be UF_6 or a mixture of UF_6, HF, and UO_2F_2. However, the chemical/thermodynamic model is invoked only when the release includes UF_6. A release of HF and UO_2F_2 is treated as a release of passive contaminants.

The model assumes that air, water vapor, and HF are ideal gases. It uses a compressibility factor to account for the deviation of UF$_6$ behavior from that of an ideal gas. Although UF$_6$ cannot exist as a liquid at atmospheric pressures, the UF$_6$ plume model includes equations for the density, vapor pressure, and enthalpy of liquid UF$_6$ because they were included in the computer code published by Williams (1985).

5.2.1 Compressibility Factor

Dewitt (1960) cites work by D.W. Magnuson in presenting the following relationship for a UF$_6$ compressibility factor:

$$Z = \frac{T_r^3}{(T_r^3 + 4.892 \times 10^5 P)},$$

(5-8)

where:

  Z = the compressibility factor (dimensionless)
  T_r = the temperature (°R), [Rankine absolute temperature]
  P = the pressure (psia)
  4.892 = a constant with the dimensions (°R^3/psia)

5.2.2 UF$_6$ Density

The density of UF$_6$ is given by the following relationships. The relationships for the UF$_6$ liquid and vapor are based on the work of Dewitt (1960), and Williams (1985) derived the relationship for the density of UF$_6$ solid based on data presented by Dewitt (1960).

The density of solid UF$_6$ is:

$$\rho_{UF6s} = 330.0 - 0.180 T_f \left(\frac{MW}{352}\right),$$

(5-9)

where:

  ρ_{UF6s} = the density of the solid UF$_6$ (lb$_m$/ft^3)
  T_f = the temperature (°F),
  MW = the molecular weight of UF$_6$

The density of liquid UF$_6$ is:

$$\rho_{UF6l} = \left(250.6 - 0.1241 T_f + 2.620 \times 10^{-4} T_f^2\right) \left(\frac{MW}{352.0}\right),$$

(5-10)

where:

  ρ_{UF6l} = the density of the liquid UF$_6$ (lb$_m$/ft^3)
  T_f = the temperature (°F),
  MW = the molecular weight of UF$_6$

The density of UF_6 vapor is:

$$\rho_{UF6v} = \frac{MW \cdot P \cdot Z}{R \cdot T_r},$$ (5-11)

where R is the universal gas constant, 10.73 (psia-ft³)/(lb-mol °R)

5.2.3 UF₆ Vapor Pressure

The following relationships, based on the work of Dewitt (1960), describe the vapor pressure of UF_6. The constants in the relationships assume English units for pressure, temperature, and volume.

From 32°F to the triple point of 147.3°F, the vapor pressure of UF_6 in the solid phase is

$$P_{UF6s} = exp\left[10.44 + 9.642 \times 10^{-3}T_f - \frac{3.90 \times 10^3}{(T_f + 298.1)}\right],$$ (5-12)

where:

P_{UF6s} = vapor pressure (psia)
T_f = temperature (°F)

From the triple point (147.3 °F) to 240 °F, the vapor pressure is:

$$P_{UF6vl} = exp\left[12.16 - \frac{4.668 \times 10^3}{(T_f + 367.5)}\right],$$ (5-13)

and from 276 °F to the critical temperature (446 °F) the vapor pressure is:

$$P_{UF6vh} = exp\left[13.76 - \frac{6.976 \times 10^3}{(T_f + 511.9)}\right]$$ (5-14)

Between 240 °F and 276 °F, the vapor pressure is estimated by a weighted average of P_{UF6vl} and P_{UF6vh}:

$$P_{UF6v} = P_{UF6vl}(276.0 - T_f) + P_{UF6vh}\left[\frac{(T_f - 240.0)}{36.0}\right]$$ (5-15)

5.2.4 UF₆ Enthalpy

Williams (1985) provides the following equations for the enthalpy of UF_6 using 25 °C (77 °F) as a base. The equations are to a large extent based on the data of Dewitt (1960).

For solid UF_6, the equation for enthalpy is:

$$H_{UF6s} = 50.446 - 5.70531 \times 10^{-2}T_r + 1.27509 \times 10^{-4}T_r^2 - 9645.63T_r^{-1},$$ (5-16)

where H_{UF6s} is the enthalpy (Btu/lb$_m$).

For liquid UF_6, the equation for enthalpy is:

$$H_{UF6l} = 30.6133 + 5.10057 \times 10^{-2}T_r + 5.13165 \times 10^{-5}T_r^2 - 6.139.34T_r^{-1}$$
$$+ 0.18268 \left[\frac{(P - P_o)}{r_l} \right] \tag{5-17}$$

where:

H_{UF6l} = the enthalpy
P = the atmospheric pressure (psia),
P_o = the vapor pressure over liquid UF_6 (psia),
ρ_l = the density of the liquid (lb_m/ft^3).

The last term in this relationship is a correction for supersaturated liquids (assuming an incompressible fluid).

Finally, the enthalpy for UF_6 vapor is:

$$H_{UF6v} = 43.2614 + 9.21307 \times 10^{-2}T_r + 6.26265 \times 10^{-6}T_r^2 + 2951.71T_r^{-1}$$
$$+ 3.0939 \times 10^{-3}T_r \left(Z|_{P,T} - Z|_{14.7,T} \right), \tag{5-18}$$

where $Z_{P,T}$ is the compressibility factor at pressure P and temperature T.

The last term in this relationship is a compressibility correction. This term is small in the atmosphere because atmospheric pressure is always near 14.7 psia.

5.2.5 Uranium Enrichment

Williams'(1985) model includes correction terms for the molecular weight to account for enrichment. The correction terms are retained in the UF_6 plume model. The molecular weight of enriched uranium is input to the model along with the release rates. RASCAL 4 corrects for molecular weight; however, the correction has only a very small effect.

5.2.6 HF-H₂O System

Williams' (1985) model treats HF and H_2O as a system for the computation of vapor pressures and enthalpy assuming that the HF and H_2O are vapors in equilibrium with a condensed phase. It is unlikely that a condensed phase will occur in the atmosphere because of the exothermic nature of the UF_6/H_2O reaction. However, the UF_6 plume model includes the equations for the condensed phase for completeness. The model assumes that HF vapor in the atmosphere exists as a set of polymers linked by hydrogen bonding. The effects of this self-association are included in the HF vapor pressure and enthalpy calculations.

5.2.7 HF Self-Association

Williams (1985) and Beckerdrite et al. (1983) report that the self-association of HF is reasonably modeled by assuming equilibrium among an HF monomer $(HF)_1$, an HF trimer $(HF)_3$, and an HF hexamer $(HF)_6$. The partial pressure of HF is:

$$P_{HF} = P_{(HF)_1} + K_3 P_{(HF)_1}^3 + K_6 P_{(HF)_1}^6, \tag{5-19}$$

where the second and third terms on the right are the partial pressures of the polymers, and K_3 and K_6 are equilibrium coefficients. Strohmeier and Briegleb determined the equilibrium coefficients experimentally (Beckerdrite et al., 1983). Using these data, Williams (1985) derived the following relationships to estimate the coefficients:

$$K_3 = exp(2.3884.0T_r^{-1} - 51.2393), \text{ and} \qquad (5\text{-}20)$$

$$K_6 = exp(40319.6T_r^{-1} - 87.7927) \qquad (5\text{-}21)$$

With self-association, the effective molecular weight for HF for vapor-phase densities and mole fractions is greater than the molecular weight of the HF monomer. It is:

$$MW_{HF} = \frac{\left[P_{(HF)_1}MW_{(HF)_1} + K_3 P_{(HF)_1}^3 MW_{(HF)_3} + K_\epsilon P_{(HF)_1}^6 MW_{(HF)_6}\right]}{P_{HF}} \qquad (5\text{-}22)$$

5.2.8 Partial Vapor Pressures of HF

If a condensed phase exists in the HF-H_2O system, relationships of the following form are used to calculate vapor pressure of HF:

$$P_{HF} = exp(AT_r^{-1} + B), \qquad (5\text{-}23)$$

where A and B are the model parameters that are functions of the weight fraction of HF in the condensed phase.

Williams (1985) gives estimates of the coefficient values based on a figure supplied by Allied Chemical Corporation (Brian C. Rogers). The differences between partial vapor pressures estimated using the model and the figure range from about 1 percent for weight fractions near 1.0 to a maximum of 5 percent at weight fractions below 0.5. If a condensed phase does not exist, the calculation of the partial vapor pressure of HF is done using an iterative procedure along with an estimation of the effective molecular weight.

5.2.9 Partial Vapor Pressure of H_2O

Until all the UF_6 has reacted with water, the UF_6/H_2O reaction will use all water entering the plume to form HF and UO_2F_2. Under these conditions, the H_2O partial vapor pressure in the HF-H_2O system will be zero. Following conversion of all of the UF_6, an initial estimate is made of the H_2O partial vapor pressure from the mass of water in the plume using the ideal gas law. The sum of the HF partial pressure and the initial estimate of the H_2O partial pressure is compared to the total pressure of HF and H_2O for an azeotropic mixture. If the sum is less than the total pressure for the azeotropic mixture, condensation does not occur, and, therefore, the initial H_2O partial pressure estimate is used. If condensation occurs, an iterative procedure is used to determine the partial pressure of H_2O. The procedure is described in detail by Williams (1985).

5.2.10 Enthalpy of HF-H_2O Vapor Mixtures

The enthalpy of HF-H_2O vapor mixtures is:

$$\begin{aligned} H_{HFH2Ov} = {}& 1051.0 + 0.472T_f \\ & - \left[376.0 + 0.136T_f + 790.642W_{(HF)_3} + 667.358W_{(HF)_6}\right]W_{HFv}, \end{aligned} \qquad (5\text{-}24)$$

where:

> $W_{(HF)3}$ and $W_{(HF)6}$ = the weight fractions of the HF polymers with respect to total HF
> W_{HFv} = the weight fraction of HF in the HF-H$_2$O vapor

The heat of association for $(HF)_3$ is -790.642 Btu/lb$_m$ of $(HF)_3$ formed, and the heat of association for $(HF)_6$ is -667.358 Btu/lb$_m$ of $(HF)_6$ formed.

5.2.11 Enthalpy of HF-H$_2$O Liquid Mixtures

A relationship of the following form gives the enthalpy of a liquid HF-H$_2$O mixture:

$$H_{HFH2Ol} = A_i + B_i W_{HFl} + C_i W_{HFl}^2,$$ (5-25)

where the coefficients A$_i$, B$_i$, and C$_i$ are functions of the weight fraction W_{HFl} of HF in the HF-H$_2$O liquid mixture. Williams (1985) provides correlations for estimating the coefficients that are based on an enthalpy-concentration diagram provided by Brian C. Rogers at Allied Chemical Corporation.

5.2.12 UO$_2$F$_2$ Enthalpy

UO$_2$F$_2$ is formed as a product of the UF$_6$-H$_2$O reaction. It is a solid with a heat capacity of 0.0821 Btu/(lb$_m$ °F). The enthalpy at any temperature relative to a reference temperature is:

$$H_{UO2F2} = 0.0821 (T_f - T_{ref}),$$ (5-26)

where:

> T_f = UO$_2$F$_2$ temperature
> T_{ref} = reference temperature (both in °F or °R).

The reference temperature is 77 degrees F in the UF$_6$ plume model.

5.2.13 Mixture Enthalpies and Plume Temperature

The UF$_6$ plume model assumes that mixing and reactions take place under constant pressure. It assumes the following reference conditions for enthalpy calculations:

- a pressure of 1013.25 mb (1 atmosphere, 760 mm Hg, or 14.696 psia)
- a temperature of 25 °C (77 °F)
- a vapor state for UF$_6$, H$_2$O, and air
- monomeric vapor for HF
- a solid state for UO$_2$F$_2$

The model calculates the enthalpy of the plume for the control volume as the control volume moves downwind. The control volume initially consists of the volume of the UF$_6$ plus the volume of the entrained air and water vapor and has an enthalpy equal to the sum of enthalpies of the UF$_6$, air, and H$_2$O. With the UF$_6$-H$_2$O reaction, the enthalpy of the control volume increases because of the heat release and changes in the masses of the plume constituents.

The UF_6-H_2O reaction is limited by one constituent or the other. If the available water is the limiting factor, the heat of reaction is:

$$H_{rxn} = 25.199 \times 10^3 \frac{m_{H2O}}{MW_{H2O}},$$

(5-27)

where:

H_{rxn} = heat of reaction (Btu)
m_{H2O} = mass of water available for the reaction (lb_m)
MW_{H2O} = molecular weight of water (lb_m/lb_m-mole)

Otherwise, the heat of reaction, which is limited by the available UF_6, is:

$$H_{rxn} = 50.398 \times 10^3 \frac{m_u}{MW_u},$$

(5-28)

where:

m_u = mass of UF_6 available for the reaction (lb_m)
MW_u = molecular weight of UF_6 (lb_m/lb_m-mole)

Note that the constants in Equations 5-27 and 5-28 have units of Btu/(lb_m-mole).

With completion of the UF_6-H_2O reaction, the enthalpy of the plume in the control volume is:

$$H_{plume} = \Delta H_{air} + \Delta H_{H2Ov} + \Delta H_{UF6} + \Delta H_{HFH2O} + \Delta H_{UO2F2} + H_{rxn}$$

(5-29)

The change in enthalpy of air is:

$$\Delta H_{air} = 0.24037 m_{air}(T_{air} - 77.0)$$

(5-30)

In addition, the change in enthalpy associated with entrained water is:

$$\Delta H_{H2O} = (0.99783 m_{H2Ole} + 0.472 m_{H2Ove})(T_{air} - 77.0),$$

(5-31)

where:

m_{H2Ole} = mass of liquid water entrained
m_{H2Ove} = mass of water vapor entrained

Finally, the model uses an iterative procedure to arrive at a plume temperature that gives the same mixture enthalpy. During this procedure, the model adjusts the phase composition of the HF-H_2O mixture and UF_6 as the temperature changes. The convergence criterion for plume temperature is 0.1 °C. This precision is more than adequate because the plume temperature is used only in plume-rise calculations.

5.3 Dispersion and Deposition of HF and UO_2F_2

The UF_6 model works in two stages. In the first stage, the model calculates the spread of UF_6, the conversion of UF_6 to HF and UO_2F_2, and the plume rise of the HF and UO_2F_2. The products of this stage are UF_6, HF, and UO_2F_2 source terms and the plume rise of HF and UO_2F_2—all as a function of distance from the release point. In the second stage, a straight-line Gaussian model (based on the model described in Chapter 4) is used to calculate airborne concentrations and deposition of HF and UO_2F_2 at receptors on a polar grid. The distance-dependent source terms calculated in the first stage are used as long as UF_6 is present. After the UF_6 is gone, the model depletes the HF and UO_2F_2 source terms to account for deposition. If UF_6 is released within a building, it uses only the second stage because of the assumption that the UF_6 has been converted to HF and UO_2F_2 within the building.

The UF_6 chemical and thermodynamics models are run in the first stage while the control volume moves downwind in small time steps. The maximum time step is 15 seconds. If, with the 1-meter wind, the UF_6 control volume reaches the first arc of receptors in less than 75 seconds, the time step is reduced so that the control volume reaches the first arc at the end of the fifth time step. As the control volume moves downwind, the stable plume equations discussed in Section 4.5.2 are used to calculate plume rise. In addition, RASCAL 4 calculates a transition plume rise (Briggs, 1984) using:

$$\Delta h_t = 1.6 \, F_b^{1/3} x^{2/3} \, u^{-1},$$

(5-32)

where:

Δh_t = transition rise (m)
F_b = buoyancy flux (m^4/s^3)
x = downwind distance
u = 10-m wind speed (m/s).

The smaller of the transition and final rise is selected as the plume rise.

The calculation of the dispersion of the HF and UO_2F_2 plumes is done using the dispersion parameters used in the main RASCAL 4 plume model. As long as unreacted UF_6 is present, the HF and UO_2F_2 plumes are assumed to be uniformly mixed in the vertical because the plumes are being fed by the UF_6-H_2O reaction. While unreacted UF_6 is present, the UF_6 model calculates normalized HF and UO_2F_2 concentrations using:

$$\chi/Q' = \frac{1}{(2\pi)^{1/2}u\Sigma_y H} exp\left[-\frac{1}{2}\left(\frac{y}{\Sigma_y}\right)^2\right],$$

(5-33)

where:

$$\Sigma_y = \left[\sigma_y^2 + \left(\frac{w_{UF6}}{4}\right)^2\right]^{1/2}, \text{ and}$$

(5-34)

$$H = \Delta h_t + 3\sigma_z$$

(5-35)

In these last two equations, σ_y and σ_z are the horizontal and vertical dispersion parameters for a point source plume (Section 4.3), and w_{UF6} is the width of the UF_6 control volume.

After all UF_6 is converted to HF and UO_2F_2, the UF_6 model calculates the normalized concentrations using:

$$\chi/Q' = \frac{1}{\pi u \Sigma_y \Sigma_z} exp\left[-\frac{1}{2}\left(\frac{y}{\Sigma_y}\right)^2\right] F(x),$$

(5-36)

where w_{UF6} is a constant equal to its value just before the last UF_6 is converted to HF and UO_2F_2, and

$$\Sigma_z = \left[\sigma_z^2 + \left(\frac{H}{2} \right)^2 \right]^{1/2} \qquad (5\text{-}37)$$

As with w_{UF6}, H is a constant equal to its value just before the last UF_6 is converted to HF and UO_2F_2. Finally, $F_x(x)$ is the vertical distribution function described in Section 4.1.1. The receptor height z is assumed to be 1 meter.

5.4 Dispersion Result Types

The RASCAL 4 UF_6 plume model calculates the following result types as a function of distance:

- Airborne Uranium Exposure ($g\text{-}s/m^3$). The airborne uranium exposure includes total exposure to uranium. It includes contributions from both UF_6 and UO_2F_2. For this calculation only, the UF_6 model assumes that UF_6 is a trace gas, not a dense gas.

- Inhaled Uranium (mg). The model calculates inhaled uranium from the total exposure using a breathing rate passed from the user interface. The default breathing rate is 3.33×10^{-4} m^3/s.

- Committed Effective Dose Equivalent (CEDE) from Inhaled Uranium (rem). The model calculates the CEDE from uranium from the inhaled uranium, the specific activity of the uranium, and the inhalation dose conversion factors from Federal Guidance Report No. 11, "Limiting Values of Radionuclide Intake and Air Concentration and Dose Conversion Factors for Inhalation, Submersion, and Ingestion." (Eckerman et al., 1988), or International Commission on Radiological Protection Publication 60 (ICRP-60), "1990 Recommendations of the International Commission on Radiological Protection," (ICRP, 1991). The Federal Guidance Report No. 11 inhalation dose conversion factors for uranium are for the D clearance class, and the ICRP-60 inhalation dose conversion factors for uranium are for the F clearance class. These clearance classes are appropriate for uranium in UF_6 and UO_2F_2 (Brodsky, 1996). All external dose calculations have been eliminated for uranium because uranium cloudshine and groundshine doses are negligible. Thus, the CEDE can be considered equivalent to the TEDE.

- Deposited Uranium (g/m^2). Deposited uranium includes uranium in any UF_6 that condenses before reacting with atmospheric water and uranium in UO_2F_2 that deposits from the UO_2F_2 plume. The UF6 model does not include UF_6 vapor deposition or enhanced deposition following the $UF_6\text{-}H_2O$ reaction. The deposition velocity for particles, such as UO_2F_2, is a function of stability and wind speed. Section 4.5.1 discusses the estimation of deposition velocities for particles, and Table 4.1 lists representative values for deposition velocities. Typical deposition velocities for particles range from about 0.3 cm/s to about 0.9 cm/s.

- Average HF Concentration in the Lung (ppm by volume). The HF concentration calculated by the UF_6 plume model is the HF concentration in the lungs. The concentration in the lungs includes inhaled HF plus HF formed as a result of the reaction of inhaled UF_6 with water in the lungs. If no UF_6 is inhaled, the HF concentration in the lung is the same as the concentration in the atmosphere.

- One-Hour Equivalent HF Concentration in the Lung (ppm by volume). The 1-hour equivalent HF concentration in the lung is an effective concentration calculated for short-duration releases for comparison to toxicity limits. The UF_6 model calculates the 1-hour equivalent HF concentration using:

$$C_{1he} = C(t)\left(\frac{t}{3600}\right)^{1/2}, \qquad\qquad (5\text{-}38)$$

where:

C_{1he} = 1-hour equivalent concentration (ppm),
$C(t)$ = average concentration for duration t (ppm),
t = duration of the exposure to concentration $C(t)$ (s).

- HF Deposition (g/m^2). The model calculates HF deposition from the atmospheric HF exposure, assuming that HF is a reactive gas and that its deposition is similar to the deposition of I_2 gas. The deposition velocity for reactive gases is a function of stability and wind speed. Section 4.5.1 discusses the estimation of deposition velocities for reactive gases, and Table 4.1 lists representative values for deposition velocities. Typical deposition velocities for reactive gases range from about 0.2 cm/s to about 1.6 cm/s.

5.5 Comparison of RASCAL UF$_6$ Plume Model with Experimental Measurements and Results from Other Models

An evaluation of the transport and dispersion portions of the UF$_6$ plume model was done for small UF$_6$ releases by comparison to measurements from three French experiments and comparison to two other UF$_6$ models. NUREG/CR-6481, "Review of Models Used for Determining Consequences of UF$_6$ Releases—Model Evaluation Report," (Nair et al., 1997), describes the experiments, data, and other models. Figure 5-1 shows the comparisons. The figure compares average uranium concentrations predicted by the RASCAL 4 UF$_6$ plume model and the other models to average concentrations measured between 10 and 500 meters from the release point. In general, the RASCAL 4 UF$_6$ plume model tends to over predict the uranium concentrations by less than a factor of 2. The other models tend to over predict by larger factors.

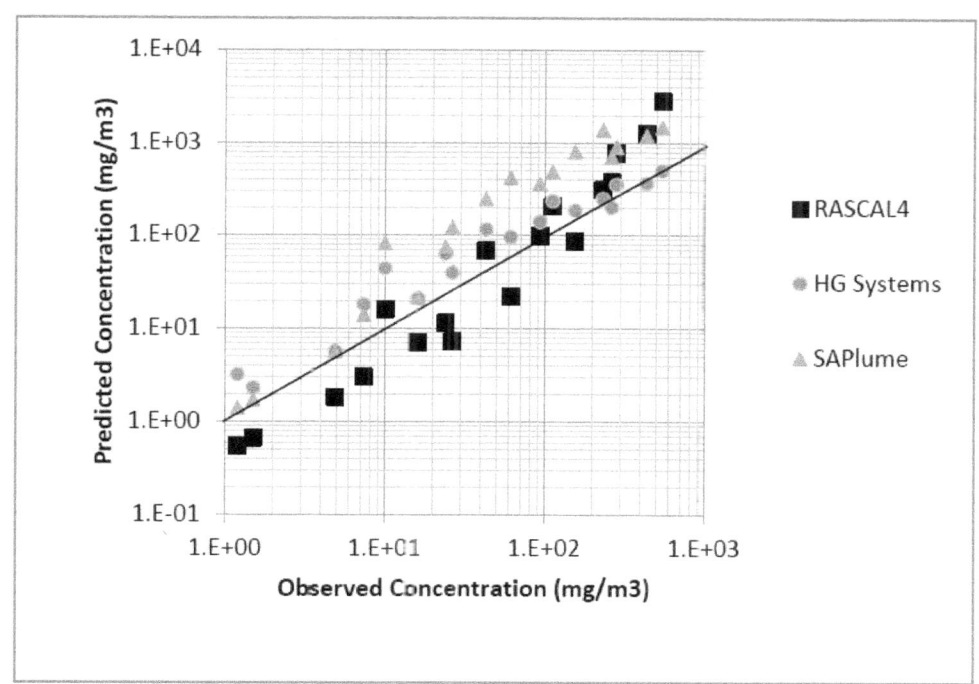

Figure 5-1 Comparison of UF$_6$ plume model predictions of average uranium concentrations to measured concentrations and predictions of other models

5.6 References

Beckerdrite, J.M., D.R. Powell, and E.T. Adams. 1983. "Self-Association of Gases, 2: The Association of Hydrogen Fluoride Data," *Journal of Chemical and Engineering,* 28:287–93.

Briggs, G.A. 1984. "Plume Rise and Buoyancy Effects," *Atmospheric Science and Power Production.* Ed., D. Randerson. DOE/TIC-27601, U.S. Department of Energy, Washington, DC.

Brodsky, A. 1996. *Review of Radiation Risks and Uranium Toxicity with Application to Decisions Associated with Decommissioning Cleanup Criteria.* RSA Publications, Hebron, CT.

Dewitt, R. 1960. "Uranium Hexafluoride: A Survey of the Physico-Chemical Properties," GAT-280. Goodyear Atomic Corporation, Portsmouth, OH.

Eckerman, K.F., A.B. Wolbarst, and A.B.C. Richardson. 1988. "Limiting Values of Radionuclide Intake and Air Concentration and Dose Conversion Factors for Inhalation, Submersion, and Ingestion," Federal Guidance Report No. 11. EPA-520/1-88-020, U.S. Environmental Protection Agency, Washington, DC.

Eidsvik, K.J. 1980. "A Model for Heavy Gas Dispersion in the Atmosphere," *Atmospheric Environment,* 14:769–777.

International Commission on Radiological Protection (ICRP). 1991. "1990 Recommendations of the International Commission on Radiological Protection," ICRP Publication 60, *Annals of the ICRP,* 21:1–3.

Nair, S.K., D.B. Chambers, S.H. Park, Z.R. Radonjic, P.T. Coutts, C.J. Lewis, J.S. Hammond, and F.O. Hoffman. 1997. "Review of Models Used for Determining Consequences of UF_6 Release—Model Evaluation Report," NUREG/CR-6481, U.S. Nuclear Regulatory Commission, Washington, DC.

Williams, W.R. 1985. "Calculational Methods for Analysis of Postulated UF_6 Releases," Volumes 1 and 2, NUREG/CR-4360, U.S. Nuclear Regulatory Commission, Washington, DC.

6. METEOROLOGICAL DATA PROCESSOR

The meteorological data processor is the part of RASCAL 4 that allows the user to enter meteorological data and prepares the data for use by the atmospheric transport and diffusion models. The user may enter meteorological data for the site (release point) and as many as 35 additional meteorological stations. All data must be entered manually at the present time. A future version of the meteorological data processor may include options for importing meteorological data files.

The following sections describe the technical aspects of the meteorological data requirements and the meteorological data processing.

6.1 Model Domain

Model domain refers to the area covered by the dose calculations in the source term to dose model. The model domain for the puff model is square. The release point is at the center of the square. The user can select a square that is 20, 50, or 100 miles (32, 80, and 160 km) on a side. The meteorological data processor creates a meteorological file for the selected domain for use by the puff model. This file describes the spatial and temporal variation of meteorological conditions at nodes on a Cartesian grid.

The meteorological data processor also creates a smaller file of meteorological data for the release point for use by the plume model and the UF_6 plume model. This file describes the temporal variation of meteorological conditions at the center of the polar grid used by these straight-line models.

The user must enter meteorological data for at least one location, generally the release site. If data are not available for the release site, the user may enter them for another location. However, the user must still select the release location as if it had meteorological data because the location of the release site is used to fix the position of the grid.

6.2 Meteorological Stations

Locations for which meteorological data are entered are called meteorological stations. Meteorological stations may be within or near the model domain. Spatial fields of winds, stability, and other parameters are produced from the station meteorological data. Section 6.3 discusses the meteorological data input.

Information on each station must be available before a station's meteorological data can be used in generating the meteorological fields needed by the puff model. RASCAL 4 includes files containing the required information for each operating nuclear power plant, large fuel cycle facilities, and several major radioactive materials facilities. Each file contains the information for the location of the site and for some selected meteorological observation stations near the site.

The meteorological data processor provides a means to add meteorological stations to the file or to update information in the file because the meteorological station file for a specific site does not include all possible meteorological stations near the site and the information for meteorological stations can become outdated.

The following information is needed for each meteorological station:

- A station identification is necessary. The station identification can be any 10-letter character. The release site identification is appropriate for the first station in the station list. Federal Aviation Administration or International Civil Aviation Organization location indicators are appropriate for the National Weather Service or other stations that have them. A longer station name may be included for each station for better identification.

- The latitude (positive north of the equator and negative south of the equator) and longitude (positive east of the prime meridian and negative west of the prime meridian) of the station in decimal degrees is necessary.

- The elevation of the station in meters above mean sea level is necessary.

- The surface roughness for the station (m) is necessary. Meteorological texts, such as Panofsky and Dutton (1984) and Stull (1988), provide guidance on estimating surface roughness. If no other information is available, a default surface roughness of 0.2 m may be used.

- The height at which the wind measurements are made (m).) is necessary. The instrument height should be the height above ground level.

The meteorological data processor uses the station position to place station data at the proper location in the modeling domain. The potential flow model that adjusts wind fields for topographic effects (Section 6.5.1) uses station elevations, and calculation of wind speed variation with height uses the surface roughness and height of wind measurements (Section 6.4.2).

The first station in the station file is the release point (the site). The latitude and longitude of the first station define the coordinates of the center of the model domain grid. The meteorological data processor calculates the distances from the center of the grid to the other stations using:

$$x_{rs} = r_e \Delta\lambda_{rs} \cos\phi_s, \text{ and} \tag{6-1}$$

$$y_{rs} = r_e \Delta\phi_{rs}, \tag{6-2}$$

where:

x_{rs} = distance of the station east (positive) or west (negative) of the source (center of the grid) (km),
y_{rs} = distance of the station north (positive) or south (negative) of the source (center of the grid) (km),
r_e = radius of the earth (6370 km),
φ_s = latitude of the center of the grid (release point)
$\Delta\lambda_{rs}$ = longitude difference between the station and the source (center of the grid) (radians)
$\Delta\varphi_{rs}$ = latitude difference between the station and the source (center of the grid) (radians)

6.3 Meteorological Data Input

The user enters station meteorological data for specific dates and times. The data may be actual observations (measurements), or they may be taken from meteorological forecasts. If available, the user should enter the following data for each station:

- Determine whether the data are observations or forecasts and enter those data.

- Enter the time of the data. The program will round the time to the nearest quarter hour. (For example, an entry of 12:07 would be changed to 12:00. Similarly, an entry of 14:22 would be changed to 14:30.)

- Enter the surface-level wind speed.

- Enter the surface-level wind direction.

- Enter the estimated atmospheric stability.

- Enter the precipitation type.

- Enter the ambient air temperature (release point only).

- Enter the estimated mixing height (optional).

If the calculation is for a UF_6 release, the user should also enter the following data for the release point (center of the grid):

- ambient air temperature
- pressure
- humidity measurement (dew point temperature, relatively humidity, or wet bulb temperature)

The thermodynamic calculations in the UF_6 plume model uses air temperature, pressure, and humidity. Only the UF_6 plume model uses pressure and humidity.

The following sections describe the meteorological variables in more detail.

6.3.1 Surface Winds

The surface winds are made up of the wind direction (the direction from which the wind is blowing) and the wind speed. The user must enter wind directions in degrees from 0 to 360 degrees. The user can enter wind speed in units of meters per second, miles per hour, or knots, and wind speed can range from 0 to 30 m/s (or equivalent in other units).

6.3.2 Atmospheric Stability Class

Atmospheric stability classes range from A (extremely unstable with rapid dispersion) through G (extremely stable with very limited dispersion). Pasquill (1961), Gifford (1961), and Turner (1964) discuss stability classes A through F as aids in estimating dispersion and relate them to observable meteorological conditions, such as solar insolation, wind speed and gustiness, and cloud cover; the stability classes themselves are not observable atmospheric variables. U.S. Nuclear Regulatory Commission (NRC) guidance (Regulatory Guide 1.23, "Onsite Meteorological Programs," Revision 1, (NRC, 2007)) recognizes the original six stability classes and an additional stability class (G) and relates the stability classes to vertical temperature gradient (*delta T* or *dT/dz*).

The user may enter atmospheric stability as either a stability class or as a vertical temperature gradient (*dT/dz*). If the user enters atmospheric stability as a vertical temperature gradient, the meteorological processor will convert the temperature gradient to a stability class using the relationship in Regulatory Guide 1.23, which is shown in Table 6-1.

Table 6-1 Estimated Pasquill-Gifford Stability Class Based on the NRC Delta T Method*

STABILITY CLASS	dT/dz (°C/100 m)	dT/dz (°F/100 ft)
A	<-1.9	<-1
B	<-1.7	<-0.9
C	<-1.5	<-0.8
D	<-0.5	<-0.3
E	< 1.5	< 0.8
F	< 4	< 2.2
G	≥ 4	≥ 2.2

*Reference: NRC, 2007.

If the user does not enter either a stability class or temperature gradient, the meteorological processor will attempt to estimate a stability class for each station from the available meteorological data. If wind speed and precipitation data are available, the meteorological processor will estimate stability class using the relationships presented in Table 6-2, which is based on a similar table presented by Turner (1969).

Earlier RASCAL versions had user options that allowed the code to set missing stability classes based on persistence and the option to allow the code to adjust stabilities for reasonableness. RASCAL 4 will always estimate missing stabilities as discussed above. In addition, RASCAL 4 does not check stabilities for reasonableness. If the user enters a stability value, the code uses that value whether it is physically reasonable or not.

Table 6-2 Estimated Atmospheric Stability Class for Missing Stability Classes

WIND SPEED (m/s)	DAY		NIGHT	
	NO PRECIPITATION	PRECIPITATION	NO PRECIPITATION	PRECIPITATION
≤2.0	B	C	F	E
2.01 to 3.0	B	C	E	E
3.01 to 5.0	C	D	E	D
≥5.01	D	D	D	D

6.3.3 Precipitation Type

Precipitation affects wet deposition in the plume model, puff model, and UF_6 plume model and the reaction between UF_6 and water in the UF_6 plume model. The user may enter information on precipitation for each station during meteorological data entry by selecting one of seven precipitation types or unknown, if appropriate. Precipitation types are:

- none
- light, moderate, and heavy rain, and

- light, moderate, and heavy snow.

Rain includes drizzle, freezing rain, and freezing drizzle. Snow includes snow grains, snow pellets, ice pellets, ice crystals, and hail. The meteorological data processor estimates precipitation rates from these precipitation types (Section 6.4.6).

6.3.4 Mixing Height

The plume and puff models use the mixing height to limit vertical dispersion. The user may enter mixing heights with the other meteorological data for a station. However, this information is generally not available. Consequently, unless the user specifically selects the option of entering mixing height data, the meteorological data processor will estimate mixing heights from wind speed and stability. An option of using climatological mixing height estimates in place of measured or calculated values also exists.

6.3.5 Temperature

The user should enter an ambient air temperature for the release point. Ambient air temperature is used to determine wet deposition rates for consequence assessments when it is snowing, and it is used in assessing the consequences of UF_6 releases. If the user omits the temperature, the meteorological data processor will use a climatological value based on location and time of year, if available. Finally, if observations do not include a temperature and if a climatological temperature is not available, then RASCAL will use a temperature of 10° C (50° F).

6.3.6 Pressure

The thermodynamic calculations in the UF_6 plume model need the station atmospheric pressure (not sea-level pressure at the station). However, the calculations are not particularly sensitive to the pressure as long as the pressure is within a few percent of the actual value. The meteorological data processor includes climatological pressures for fuel cycle facilities that should be adequate for most purposes because atmospheric pressures rarely vary by more than ±5 percent. If there is neither an observed pressure nor a climatological pressure, RASCAL 4 assumes a default pressure of 950 mb, which is equivalent to an elevation of about 550 m (1800 ft) msl in the standard atmosphere. The program will convert pressures entered in other units to millibars.

6.3.7 Humidity

Chemical reaction and thermodynamic calculations in the UF_6 plume model need information on humidity. The user may enter humidity information for the release point as dew point temperature, relative humidity, or wet bulb temperature. The meteorological data processor includes humidity information based on climatological data for fuel cycle facilities. However, the user should enter actual data whenever possible because the calculations are very sensitive to humidity, and humidity has a wide range of variation in the atmosphere. If the user does not enter humidity information and if climatological data on humidity is not available, RASCAL 4 uses a default relative humidity of 60 percent unless precipitation occurs. If there is precipitation, RASCAL 4 assumes that the relative humidity is 95 percent.

6.3.8 Temporal Interpolation of Input Values

The atmospheric models in RASCAL 4 expect meteorological data on the 15-minute time interval typically used to record meteorological data at U.S. nuclear power plants. Data from other meteorological stations are not likely to be available on that interval. Consequently, the meteorological data processor will estimate missing 15-minute data for each station by linear interpolation between entered values. For example, if 10:00 and 11:00 observations are entered for a station, the program will estimate values for 10:15, 10:30, and 10:45. Table 6-3 lists the rules and methods of interpolation.

Table 6-3   Interpolation Rules

CONDITION	HOW IT IS HANDLED
Observation to observation	Interpolate all times between.
Observation to forecast or forecast to forecast	Interpolate if time difference is 30 minutes or less; otherwise, persist for half the time interval, and then interpolate the remainder.
Time steps before the first defined data (observation or forecast)	Set all times preceding the defined data as missing.
Time steps from last observation or forecast to last time step	Persist for up to 2 hours, and then set times as missing if needed.

6.3.8.1 Wind Interpolation Method

If the winds for both the earlier and later observation are valid, the winds are interpolated as follows:

- The wind speed and direction are converted to U (east-west) and V (north-south) components.

- The U and V components are linearly interpolated (i.e., $U(t) = (U_l - U_e)\{(t-t_e)/(t_l-t_e)\} + U_e$, where U_l, U_e, t_l, and t_e are the U component and time of the later and earlier observations, respectively).

- The U and V components are converted back to speed and direction.

6.3.8.2 Atmospheric Stability Interpolation Method

The Pasquill-Gifford stability class is converted to a numerical value (1-7). Estimation of the atmospheric stability is done using linear interpolation between the two observations. If the interpolated values is not an integer, then it is rounded to the nearest integer.

6.3.8.3 Precipitation-Type Interpolation Method

The precipitation type for the earlier entry is used when the time is less than, or equal to, halfway between the two observations. If the time is greater than halfway between the two observations, the precipitation type of the later observation is used.

6.3.8.4 Mixing Height Interpolation Method

Unless the mixing height is being entered directly (not calculated from the meteorological data or from climatology), the method of estimating mixing heights that is being used for the earlier observation will continue to be used.

If the mixing heights are being entered directly, the technique used to interpolate stability is used to estimate the missing mixing heights (see previous statement).

6.3.8.5 Temperature, Pressure, and Moisture Interpolation Method

The same technique previously explained to estimate missing stabilities is used for temperature, pressure, and humidity.

6.4 Other Meteorological Parameters

Meteorological data entered for a station are used to evaluate additional parameters. The following subsections describe these additional parameters.

6.4.1 Monin-Obukhov Length

The Monin-Obukhov length (L) is a scaling length for vertical motions in atmospheric boundary layer studies that is used as a measure of atmospheric stability. It is used in wind profile, turbulence, and mixing layer depth calculations. Golder (1972) provides a graphical means for converting from Pasquill-Gifford stability classes to Monin-Obukhov lengths using the surface roughness length (Section 6.4.2). The meteorological data processor uses a procedure to convert stability classes to Monin-Obukhov lengths that was developed by Ramsdell et al. (1994) based on Golder's work.

6.4.2 Wind Speed versus Height

The RASCAL 4 atmospheric dispersion models use winds that are representative of 10 meters above ground level for ground-level release calculations and winds that are representative of the release height for elevated release calculations. Wind measurements are not always made at these heights. Therefore, the meteorological data processor adjusts wind speeds for the difference between the measurement height and the height required for model calculations when the observed wind speed is 0.223 m/s or greater. A diabatic wind profile model, which accounts for the effects of surface roughness and atmospheric stability on the variation of wind speed with height, is used for this adjustment. No attempt is made to model the variation of wind direction with height.

The diabatic profile model is derived from atmospheric boundary layer similarity theory proposed by Monin and Obukhov (1954). The basic hypothesis of similarity theory is that a number of parameters in the atmospheric layer near the ground, including wind profiles, should be universal functions of the friction velocity, a length scale, and the height above the ground. The length scale is referred to as the Monin-Obukhov length, and the ratio z/L is related to atmospheric stability.

The diabatic wind profile is:

$$u(z) = \frac{u_*}{k}\left[ln\left(\frac{z}{z_0}\right) - \psi\left(\frac{z}{L}\right)\right],\tag{6-3}$$

where:

$u(z)$ = wind speed at height z (m/s),
$u*$ = friction velocity (boundary layer scaling velocity) (m/s),
k = von Karman constant (≈ 0.4)
z_0 = surface roughness length (m),
$\psi(z/L)$ = stability correction factor
L = Monin-Obukhov length (m).

The surface roughness length is associated with small-scale topographic features. It arises as a constant of integration in the derivation of the wind profile equations and is used in several boundary layer relationships. Texts on atmospheric diffusion, air pollution, and boundary layer meteorology (Panofsky and Dutton, 1984; Stull, 1988) contain tables that give approximate relationships between surface roughness and land use, vegetation type, and topographic roughness.

The term $\Psi(z/L)$ accounts for the effects of stability on the wind profile. In stable atmospheric conditions, $\Psi(z/L)$ has the form $-\alpha z/L$, where α has a value of 5. In neutral conditions, $\Psi(z/L)$ is equal to zero, and the diabatic profile simplifies to a logarithmic profile.

In unstable air, $\Psi(z/L)$ is more complicated. According to Panofsky and Dutton (1984), the most common form of $\Psi(z/L)$ for unstable conditions, based on the work of Businger et al. (Paulson, 1970), is:

$$\psi\left(\frac{z}{L}\right) = ln\left(\left[\frac{(1+x^2)}{2}\right]\left[\frac{(1+x)}{2}\right]^2\right) - 2\tan^{-1}(x) + \frac{\pi}{2},$$ (6-4)

where x is $(1-16z/L)^{1/4}$.

Equation 6-4 is used to estimate the friction velocity ($u*$) from the wind speed, surface roughness, and Monin-Obukhov length. In unstable and neutral conditions, the use of Equation 6-4 is limited to the lowest 100 meters of the atmosphere. In stable conditions, the upper limit for application of Equation 6-4 is the smaller of 100 meters or three times the Monin-Obukhov length.

The assumption is made that wind speed above 100 meters is equal to the wind speed at 100 meters. If the mixing layer thickness is less than 100 meters, the assumption is made that the wind speed above the top of the mixing layer is equal to the wind speed at the top of the mixing layer.

6.4.3 Mixing Height

Heating of the surface and surface friction combine to generate turbulence that mixes material released at or near ground level through a layer that varies in thickness from a few meters to a few kilometers. This layer is referred to as the mixing layer. The atmospheric models in RASCAL 4 use the mixing height (also referred to as the mixing layer depth and mixing layer thickness) to limit vertical diffusion.

The meteorological data processor has three methods for obtaining estimates of the mixing height at meteorological stations. The mixing height may be entered directly, or the program may estimate it from either current meteorological data or climatological information. Of the latter two options, estimation of mixing height from current meteorological data is preferable to estimating the mixing height from climatological data, if sufficient data are available.

The meteorological data processor uses algorithms developed by Ramsdell et al. (1994) to estimate mixing height from current meteorological data. The algorithms are based on relationships derived by Zilitinkevich (1972) for stable and neutral conditions.

For stable atmospheric conditions, the relationship is:

$$H = k\left(\frac{u_*L}{f}\right)^{1/2},$$ (6-5)

where:

6-8

H = mixing height (m),
k = von Karman constant (0.4)
u_* = friction velocity (m/s),
L = Monin-Obukhov length (m),
f = Coriolis parameter (1/s)

A 50-meter mixing height is used if the mixing height calculated by Equation 6-5 is less than 50 meters. Similarly, if the calculated mixing height is greater than 2,000 meters, the mixing height is set to 2,000 meters.

For neutral and unstable conditions, the mixing height is:

$$H = \frac{\beta u_*}{f},$$

(6-6)

where β is a constant set to 0.2.

If the mixing height calculated by Equation 6-6 is less than 250 meters, the mixing height is set to 250 meters, and if the calculated mixing height is greater than 2,000 meters, the mixing height is set to 2,000 meters.

Equations 6-5 and 6-6 were developed for mid-latitude locations. They include the Coriolis parameter in the denominator. The Coriolis parameter is a function of the sine of the site latitude. In latitudes above 20 degrees, these relationships should give reasonable mixing heights. However, the Coriolis parameter approaches zero as the latitude approaches the equator, and it is zero at the equator. At the equator, the mixing heights defined by Equations 6-5 and 6-6 become undefined. Consequently, near the equator RASCAL 4 may tend to underestimate doses at longer distances because the mixing height is overestimated.

Table 6-4 illustrates the variation of mixing height estimated by Equations 6-5 and 6-6 with wind speed, stability, and latitude. The Turkey Point Nuclear Generating Station and Columbia Generating Station are the southern- and northern-most nuclear power plant sites, respectively, in the United States. Note that some of the stability and wind speed combinations shown in the table are unrealistic.

Table 6-4   RASCAL 4 Mixing Height Estimates*

WIND SPEED (m/s)	TURKEY POINT GENERATING STATION LATITUDE 25.43° N. STABILITY CLASS						
	G	F	E	D	C	B	A
2	50	106	204	292	847	1,584	2,000
5	64	168	332	730	2,000	2,000	2,000
8	81	212	408	1,168	2,000	2,000	2,000
12	99	260	499	1,752	2,000	2,000	2,000
16	115	300	576	2,000	2,000	2,000	2,000
20	128	336	644	2,000	2,000	2,000	2,000
30			798	2,000	2,000	2,000	2,000

WIND SPEED (m/s)	COLUMBIA GENERATING STATION LATITUDE 46.47° N. STABILITY CLASS						
	G	F	E	D	C	B	A
2	50	82	157	250	386	722	1,333
5	50	129	248	432	1,527	2,000	2,000
8	62	163	314	692	2,000	2,000	2,000
12	77	200	384	1,038	2,000	2,000	2,000
16	88	231	444	1,384	2,000	2,000	2,000
20	99	258	496	1,730	2,000	2,000	2,000
30			607	2,000	2,000	2,000	2,000

*In meters.

Estimation of the mixing layer thickness may also be done using climatological data for sites in the RASCAL 4 facility database. When the user selects this option, the mixing layer thickness is estimated from typical morning and afternoon thicknesses for each month using the method used in the U.S. Environmental Protection Agency's (EPA's) meteorological preprocessor code, PCRAMMET (EPA, 1999). Calculation of the monthly morning and afternoon mixing layer thicknesses was done from daily data obtained from the EPA's Support Center for Regulatory Air Models (www.epa.gov/scram001/). The following rules are used to estimate mixing layer thicknesses from the monthly values:

- From midnight to sunrise, use the morning mixing height.

- From sunrise to 1400, linearly interpolate between morning and afternoon mixing heights.

- From 1400 to sunset, use the afternoon mixing height.

- From sunset to midnight, use exponential interpolation between the afternoon and morning mixing heights. For the last day of the month, use the morning of the next month.

The exponential interpolation of the mixing height is:

$$H(t) = a \cdot exp\left(-\frac{bt}{24}\right),$$
(6-7)

where:

$H(t)$ = mixing height (m) at time t
$a = H_{morn}/exp(-b)$
$b = 24 \ln(H_{aft}/H_{morn})(24 - t_{sunset})$
H_{morn} = morning mixing height (m)
H_{aft} = afternoon mixing height (m)
t_{sunset} = time of sunset (h)

The following set of equations (Stull, 1988) is used to calculate sunset and sunrise times. The first the equation calculates elevation angle of the sun. It is:

$$\sin \upsilon = \sin \phi \sin \delta_s - \cos \phi \cos \delta_s \cos(T_0)$$
(6-8)

where:

υ = local elevation of the sun
φ = latitude of the station
δ_s = solar declination angle (angle of the sun above the equator)
T_0 = local time

The second equation is used to calculate the solar declination angle. It is:

$$\delta_s = \phi_r \cos\left(\frac{2\pi(d-d_r)}{d_y}\right),$$
(6-9)

where:

φ_r = latitude of the Tropic of Cancer (23.45°),
d = Julian calendar day of the year
d_r = Julian calendar day of the summer solstice (173)
d_y = average number of days per year (365.25)

Finally, the local time is defined as follows:

$$T_0 = \left(\frac{\pi t_{utc}}{12}\right) - \lambda_e,$$
(6-10)

where

t_{utc} = time at the prime meridian

λ_e = longitude (in radians) of the station

Sunrise and sunset are calculated by setting the solar elevation angle to -0.833° (the sun appears to rise and set when it is 0.833° below the horizon) and by solving for T_0 using Equation 6-10. Sunset is 24 hours − T_0. The equations for sunrise and sunset do not take into account the ellipticity of the Earth's orbit, but it is accurate to about ±16 minutes.

Table 6-5 shows examples of the climatological mixing height estimates for Columbia Generating Station. Figure 6-1 illustrates the January and July hourly variations of mixing height estimated by RASCAL 4 from the climatological data.

Table 6-5 An example of the Climatological Mixing Heights

COLUMBIA GENERATING STATION MIXING HEIGHTS (m)		
Month	a.m.	p.m.
January	180	350
April	400	1,400
July	190	2,300
October	200	760

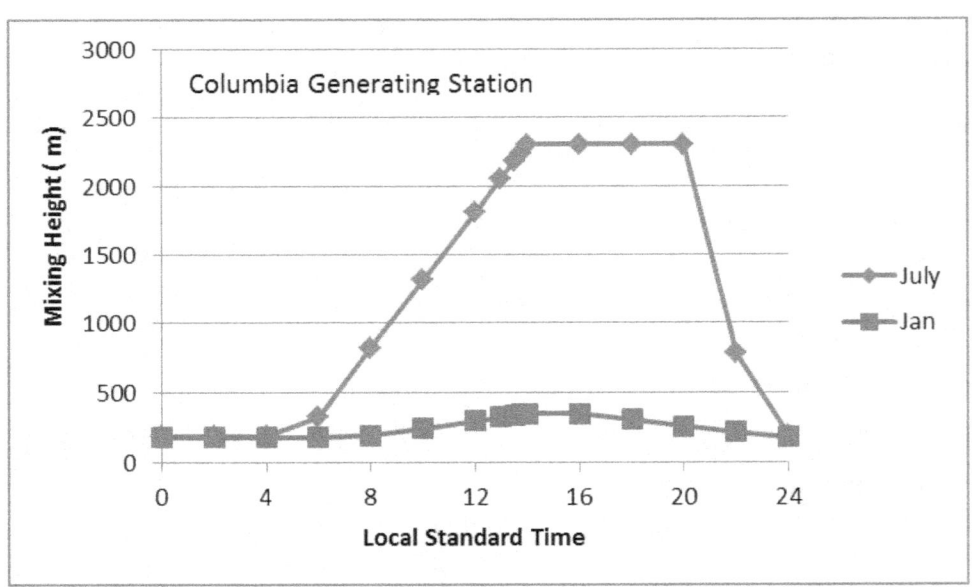

Figure 6-1 RASCAL 4 estimates of January and July mixing heights at Columbia Generating Station based on climatological data

6.4.4 Dry Air and Water Vapor Density

The UF_6 plume model requires estimates of the air density and water content of the atmosphere. Estimation of these two variables is done from the air temperature, station pressure, and humidity for the release location. (The model assumes that the release location is located at the first meteorological station. The method used to calculate the dry air density and water vapor density depends on the variables used for humidity and on whether the temperature, pressure, and moisture variables exist.

The following equation gives the water vapor density:

$$\rho_v = \frac{e}{R_v T},$$ (6-11)

where:

ρ_v = water vapor density (kg/m^3),
e = vapor pressure (Pa),
R_v = gas constant is 461.5 J/kg °K
T = temperature (°K),

The following equation is use to estimate dry air density:

$$\rho_d = \frac{(p-e)}{RT},$$ (6-12)

where:

ρ_d = dry air density (kg/m^3),
p = total station pressure (Pa),
R = gas constant for dry air is 287.0 J/kg °K

The method used to calculate the vapor pressure e depends upon the moisture variable. If the moisture variable is the dew point and if precipitation is not occurring, the following equation gives the vapor pressure:

$$e = e_s(T_d),$$ (6-13)

where $e_s(T_d)$ (mb) is the saturation vapor pressure at temperature T_d (°C).

According to Rogers and Yau (1989), the saturation vapor pressure is:

$$e_s(T) = 6.112 exp\left(17.67\frac{T}{T + 243.5}\right)$$ (6-14)

If the moisture is defined by the relative humidity and if precipitation is not occurring, the following equation gives the vapor pressure:

$$e = \frac{RH \cdot e_s(T)}{100},$$ (6-15)

where:

RH = relative humidity (percent)
$e_s(T)$ = saturation vapor pressure

If the moisture is given by the wet bulb temperature and if precipitation is not occurring, the following equation gives the vapor pressure:

$$e = e_s(T) - \left(\frac{p}{0.622}\right) \cdot \left[\frac{1004(T - T_w)}{2.5E6}\right], \tag{6-16}$$

where T_w is the wet bulb temperature (°C)

If precipitation is occurring, RASCAL 4 assumes that the air is 95 percent saturated; therefore, the vapor pressure is:

$$e = 0.95 e_s(T), \tag{6-17}$$

where Equation 6-14 gives $e_s(T)$.

During precipitation, Equation 6-17 is used regardless of the moisture variable.

If a temperature, pressure, or moisture variable is missing from the observation, RASCAL 4 uses the best available data for estimating dry air and water vapor density. If either a temperature or moisture variable is missing, the code uses a climatological or default value in its place to calculate dry air and water vapor densities. If both the temperature and the moisture variables are missing and if climatological values for dry air and water vapor density exist, the code will use those values unless precipitation is occurring. If there is precipitation, the code will increase the climatological value of water vapor density by a factor of 1.58, which is the ratio of 95 to 60 percent. Finally, if a climatological value for neither variable exists, then RASCAL 4 uses a dry air density of 1.2 kg/m^3 and a water vapor density of 5.6 g/m^3 unless precipitation is occurring. If precipitation is occurring, the code assumes that the water vapor density is 8.9 g/m^3.

6.4.5 Precipitation Rate

The RASCAL 4 atmospheric codes use precipitation rate to calculate the wet deposition rate for UF$_6$ releases. When the precipitation type for a station is other than none or unknown, the meteorological data processor estimates a precipitation rate (mm/h) for the station using the precipitation type and a precipitation rate zone. The RASCAL 4 database assigns each one of three precipitation rate zones to each site. The precipitation zones, originally defined in Ramsdell et al. (1994), are based on annual precipitation. Zone 1 is for areas in which the annual precipitation is less than 10 inches, Zone 2 is for areas in which the annual precipitation is between 10 and 20 inches, and Zone 3 is for areas in which the annual precipitation exceeds 20 inches. Most existing reactor sites are assigned to precipitation rate Zone 3. Some sites in the drier regions of the United States are assigned to Zone 1. Precipitation rate zone assignments are made in the climatology database and can be modified as appropriate.

Table 6-6 lists the precipitation rates assigned by the meteorological data processor. These rates, based on data collected in the Pacific northwest, should be conservative for most nuclear facilities in the United States.

Table 6-6   Precipitation Rates as a Function of Precipitation Climate Zone

PRECIPITATION TYPE	PRECIPITATION RATE (mm/h)		
	ZONE 1	ZONE 2	ZONE 3
Light rain	0.4	0.6	0.7
Medium rain	3.8	3.8	3.8
Heavy rain	3.8	3.8	8.5
Light snow	0.3	0.3	0.7
Medium snow	1.7	1.7	3.8
Heavy snow	1.7	1.7	3.8

6.5 Calculating Spatially Varying Meteorological Conditions

The puff model takes into account both spatial and temporal variations in the atmospheric conditions. The meteorological data processor provides the gridded fields of the atmospheric stability class, the inverse Monin-Obukhov length, the east-west (U) and north-south (V) components of the wind, the mixing height, and the precipitation type and precipitation rate for each of the three puff model domains. The following subsections describe the preparation of the fields from the station data.

6.5.1 Wind Fields

The puff model uses winds at the position of the center of the puff determined from wind fields to calculate the movement of puffs. Spacing between the wind grid points in the fields varies with the model domain. It is 1 mile for the 10-mile domain, 2.5 miles for the 25-mile domain, and 5 miles for the 50-mile domain. The initial wind fields, which consist of U and V components of the wind vectors, are estimated initially from station meteorological data using a $1/r^2$ interpolation scheme, where r is the distance from the grid point to the station. This interpolation scheme, which was used in earlier NRC codes, such as MESOI (Ramsdell et al., 1983) and MESORAD (Scherpelz et al., 1986; Ramsdell et al., 1988), is common in spatial interpolation of the wind fields (Hanna et al., 1982).

6.5.2 Adjustment of Wind Fields for Topography

If the meteorological stations reporting data are well placed with respect to major topographic features, the wind fields developed by interpolation will give reasonable puff trajectories. However, with one meteorological station or a small number of stations, the wind fields may not properly reflect the effects of topography. The meteorological data processor uses a simple one-layer model to adjust wind fields for topography. Wind field adjustments are greatest for stable atmospheric conditions (E, F, and G stability classes) and least for neutral conditions (stability class D). The meteorological data processor does not adjust wind fields in unstable atmospheric conditions (stability classes A, B, and C). For this purpose only, RASCAL assumes that the atmospheric stability at the release point (center of the model domain) applies to the entire domain. RASCAL 3 included this model as an option; it is a standard feature in RASCAL 4. The only change in the model is that in RASCAL 4 does not adjust wind fields for topography if all observed wind speeds are less than 0.223 m/s.

The wind field model in RASCAL 4 is a two-dimensional adaption of the wind-fitting program described by Ross (1988) that is used in the NUATMOS and MATTHEW codes. RASCAL 4 uses wind fields created by interpolation as the starting point in the adjustment process. The code calculates the thickness of the mixing layer for each node in the model domain by computing the difference between the top of the boundary layer and the terrain elevation. For those nodes where the terrain rises above the top of the boundary layer, the program assumes that the boundary layer thickness is 0.01 meter and sets the wind to zero. This technique is simple to implement and has proven effective at generating flows that avoid obstacles, such as mountain ridges.

RASCAL 4 then adjusts the initial wind field using methods of variational calculus to produce a nondivergent wind field in the boundary layer that is subject to the constraint of minimum difference between the initial wind field and the adjusted wind field. The procedure for adjusting the wind field involves solving Poisson's equation. The code uses a nine-point Laplacian operator and a simultaneous relaxation technique to obtain the solution.

The model has been tested and shows that the winds that are produced by the model flow around obstacles that are well resolved by the grid. Obstacles that have a width of 3 grid points or greater are considered well resolved. Smaller obstacles may or may not be resolved, depending on their shape and orientation relative to the grid. For example, a ridge that is 1 grid point wide is well resolved if it runs in the x or y direction, but if that same ridge is at 45 degrees to the grid, it is not resolved.

The adjusted wind field is most accurate near stations and along trajectories that pass near stations. Wind fields are less accurate elsewhere. Thus, having wind data near the release point and, if possible, at downwind locations is desirable.

RASCAL 4 includes topographic data for all sites in the facility database. Sites that are not included in the RASCAL database and the generic sites do not have topographic data files. Therefore, RASCAL 4 does not modify wind fields for topographic effects for these sites. Earlier RASCAL versions allowed the user to turn off the use of topography. RASCAL 4 always adjusts the wind field for topography if the data are available.

6.5.3 Stability and Precipitation

The stability class and precipitation fields (precipitation type and precipitation rate) are based on data for the closest meteorological station. Fields created in this manner include stability class, inverse Monin-Obukhov length, precipitation type, and precipitation rate. This procedure avoids averaging that would minimize the effects of extreme stability or instability. It also provides maximum detail in treating isolated precipitation events.

6.5.4 Mixing Height

The spatial variation of the mixing height is modeled in a two-step process. The first step is to create an initial mixing height field using the mixing heights at the station closest to each point in the field. If only one station exists, the process is terminated after this step. If two or more stations exist, unnatural discontinuities in the initial mixing height field are likely to exist. The second step in preparing the final mixing height field is to smooth out the field using a 25-point spatial filter. This step replaces the initial mixing height estimates by an average of 25 mixing heights (the point and 24 surrounding points). For points near the edge of the domain, RASCAL 4 assumes that the mixing heights at the domain boundary are constant outside the domain.

6.6 Calculating Meteorological Conditions at the Source

All RASCAL 4 atmospheric dispersion models require information about the wind speed, wind direction, atmospheric stability, precipitation type, precipitation rate, mixing layer depth, and temperature at the source. If these meteorological data are available for the release point, which is considered to be at the source, RASCAL 4 uses those data. If no data are available for the release point, the code will estimate the wind speed, wind direction, atmospheric stability, current weather, precipitation rate, and mixing layer depth from the spatial meteorological data field. For the temperature, the code will use a default climatological value if it exists. The default climatological temperature varies by month and is obtained from the climate file. If the climate file does not exist for the site, the code will flag the temperature as a missing value.

6.7 References

Gifford, F.A. 1961. "Use of Routine Meteorological Observations for Estimating Atmospheric Dispersion," *Nuclear Safety*, 2(4):47–51.

Golder, D. 1972. "Relations among Stability Parameters in the Surface Layer." *Boundary-Layer Meteorology*, 3(1):47–58.

Hanna, S.R., G.A. Briggs, and R.P. Hosker. 1982. *Handbook on Atmospheric Diffusion,* DOE/TIC-11223, U.S. Department of Energy. Washington, DC.

Monin, A.S., and A.M. Obukhov. 1954. "Basic Laws of Turbulent Mixing in the Ground Layer of the Atmosphere." *Tr. Akad.. Nauk SSSR Geophiz. Inst.* , 24(151):163–87. Translation available at http://gronourson.free.fr/IRSN/Balagan/Monin_and_Obukhov_1954.pdf.

Panofsky, H.A., and J.A. Dutton. 1984. *Atmospheric Turbulence,* J. Wiley & Sons, New York, NY.

Pasquill, F. 1961. "The Estimation of the Dispersion of Windborne Material," *The Meteorological Magazine,* 90:33–49.

Paulson, C.A. 1970. "The Mathematical Representation of Wind Speed and Temperature Profiles in the Unstable Atmospheric Surface Layer," *Journal of Applied Meteorology,* 9:1884–89.

Ramsdell, Jr., J.V., G.F. Athey, and C.S. Glantz. 1983. "MESOI Version 2.0: An Interactive Mesoscale Lagrangian Puff Dispersion Model With Deposition and Decay," NUREG/CR-3344, U.S. Nuclear Regulatory Commission, Washington, DC.

Ramsdell, Jr., J.V., G.F. Athey, T.J. Bander, and R.I. Scherpelz. 1988. "The MESORAD Dose Assessment Model, Volume 2: 'Computer Code,'" NUREG/CR-4000, U.S. Nuclear Regulatory Commission, Washington, DC.

Ramsdell, Jr., J.V., C.A. Simonen, and K.W. Burk. *Regional Atmospheric Transport Code for Hanford Emission Tracking (RATCHET),* PNWD-2224 HEDR, Pacific Northwest National Laboratory, Richland, WA, 1994.

Rogers, R.R., and M.K. Yau. 1989. *A Short Course in Cloud Physics,* Pergamon Press, New York, NY.

Ross, D.G. 1988. "Diagnostic Wind Field Modeling for Complex Terrain: Model Development and Testing," *Journal of Applied Meteorology,* 27:785–96.

Scherpelz, R.I., T.J. Bander, G.F. Athey, and J.V. Ramsdell. 1986. "The MESORAD Dose Assessment Model: Volume 1, 'Technical Basis,'" NUREG/CR-4000, U.S. Nuclear Regulatory Commission.

Snedecor, G.W., and W.G. Cochran. 1980. *Statistical Methods*, Seventh Edition, Iowa State University Press, Ames, IA.

Stull, R.B. 1988. *An Introduction to Boundary Layer Meteorology,* Kluwer Academic Publishers, Dordrecht, Netherlands.

Turner, D.B. 1964, "A Diffusion Model for an Urban Area," *Journal of Applied Meteorology,* 3(1):83–91.

Turner, D.B. 1969. "Workbook of Atmospheric Dispersion Estimates," Report No. 999-AP-26, U.S. Environmental Protection Agency, Washington, DC.

U.S. Environmental Protection Agency (EPA). 1995. "PCRAMMET User's Guide," EPA-454/B-96-001, U.S. Environmental Protection Agency, Washington, DC.

U.S. Environmental Protection Agency (EPA). 1995. "PCRAMMET User's Guide," U.S. Environmental Protection Agency, Washington, DC.

U.S. Nuclear Regulatory Commission (NRC). 2007. "Onsite Meteorological Programs," Regulatory Guide 1.23, Revision 1, U.S. Nuclear Regulatory Commission, Washington, DC.

Zilitinkevich, S.S. 1972. "On the Determination of the Height of the Ekman Boundary Layer," *Boundary-Layer Meteorology*, 3(2):141–5.

7. FIELD MEASUREMENT TO DOSE MODEL CALCULATIONS

Chapter 7 describes the "Field Measurement to Dose" (FMDose) module of RASCAL 4. The module, which calculates doses from field measurements of radionuclide air concentrations or surface contamination, calculates both early-phase doses and intermediate-phase doses. The updates of this module include a new resuspension model (Maxwell and Anspaugh, 2011) and an option to use International Commission on Radiological Protection (ICRP) Publication 60, "1990 Recommendations of the International Commission on Radiological Protection," issued 1991 (ICRP-60), inhalation dose conversion factors (ICRP, 1991). In addition, the early-phase calculations have been expanded to cover a 96-hour period. The calculation of early-phase doses is done from either the average air concentration during plume passage or the surface concentration immediately following plume passage. The calculation of intermediate-phase doses is done from surface contamination, which is either measured or inferred from the average air concentration during plume passage. The following sections describe the assumptions, models, and methods used in the dose computations.

7.1 Early-Phase Dose Calculations

Early-phase dose calculations may be made from either an average air concentration measured during plume passage or a ground concentration measurement made immediately after plume passage. These dose calculations assume that the measurement represents the radiological conditions during plume passage and that an individual is exposed to the plume during its passage and to radiation from the surface for the period of plume passage plus the remainder of a 96-hour period. The calculation assumes that the plume is present at the measurement location for the duration of the release and that the concentration in the plume is constant during plume passage.

7.1.1 Air Concentration Measurements

When air concentration measurements are used to estimate early-phase doses, the dose estimates for the period of plume passage have three components: (1) a submersion dose, (2) an inhalation dose, and (3) a groundshine dose. The early-phase doses for the period following plume passage also have three components: (1) a groundshine dose from surface contamination, (2) an inhalation dose from resuspended activity, and (3) a submersion doses from resuspended activity. The submersion dose from resuspended activity is generally negligible.

7.1.1.1 Plume Passage Doses

FMDose estimates the air submersion doses during plume passage using:

$$D_{asi}(\Delta T) = C_{ai} \times DCF_{asi} \times \Delta T, \tag{7-1}$$

where:

$D_{asi}(\Delta T)$ = the external dose equivalent from submersion in air concentration C_{ai} of radionuclide i for exposure period ΔT (rem)
C_{ai} = the measured average air concentration of radionuclide i, (Ci/m^3),
DCF_{asi} = the dose conversion factor for effective dose equivalent for air submersion in radionuclide i ($rem/(Ci\text{-}s\ m^{-3})$)
ΔT = the exposure period (s).

FMDose estimates of committed doses from inhalation during plume passage :

$$D_{ainhio}(\Delta T) = C_{ai} \times B \times DCF_{inhio} \times \Delta T, \qquad (7\text{-}2)$$

where:

$D_{ainhio}(\Delta T)$ = the committed effective dose equivalent (CEDE) or committed dose equivalent to organ o from inhalation of radionuclide i over period ΔT (rem)
B = the breathing rate (3.334×10^{-4} m³/s)
DCF_{inhio} = the appropriate dose conversion factor for inhaled radionuclide i (rem/Ci)

Finally, FMDose infers the groundshine doses from deposited material from the air concentration using:

$$D_{gsio}(\Delta T) = C_{ai} \times v_d \times DCF_{gsio} \times \frac{\Delta T^2}{2}, \qquad (7\text{-}3)$$

where:

$D_{gsio}(\Delta T)$ = the effective dose equivalent to organ o from groundshine from radionuclide i over the period of plume passage ΔT (rem)
v_d = the deposition velocity (m/s),
DCF_{gsio} = the appropriate dose conversion factor for external exposure groundshine from radionuclide i

A constant deposition velocity of 0.005 m/s is used to infer the surface contamination. This deposition velocity, which represents meteorological conditions of D stability and a wind speed of about 2 m/s, is used for both iodines and other particles. If the meteorological conditions during the release are known, the deposition velocities in Table 4-1 can be used to adjust early groundshine doses based on an average air concentration measurement for different meteorological conditions.

The dose conversion factors for early-phase doses are taken from four sources. Dose conversion factors for cloud submersion doses are from Federal Guidance Report No. 12, "External Exposure to Radionuclides in Air, Water, and Soil," (Eckerman and Ryman, 1993) as updated in DCFPAK2 (Eckerman et al., 2008). Users may select dose conversion factors for CEDE and thyroid dose from either FGR 11, "Limiting Values of Radionuclide Intake and Air Concentration and Dose Conversion Factors for Inhalation, Submersion, and Ingestion," (Eckerman et al., 1988), or from ICRP-60 (ICRP, 1991). Bone inhalation, colon, and lung dose conversion factors are acute dose conversion factors from the International Atomic Energy Agency (IAEA, 2006).

RASCAL 4 sums early-phase doses over all radionuclides entered by the user. In general, these doses do not include either decay or ingrowth and decay of progeny. However, if a nuclide has a short-lived daughter (<15 minutes), RASCAL 4 treats the parent-daughter combination as a single nuclide with the parent's half-life. Nuclides with an implicit daughter are indicated by an * or + symbol following the nuclide identification. The code assumes that the implicit daughter is in radiological equilibrium with the parent. The effective dose conversion factor for the combination of the parent and daughter is the sum of the parent's effective dose rate coefficient plus the daughter's effective dose rate coefficient (corrected for a branching ratio). Appendix A discusses the treatment of these combinations of parent plus short-lived daughter in more detail; Table A-1 lists the nuclides with implicit daughters.

7.1.1.2 Postplume Doses

When early-phase doses are calculated from an average air concentration, RASCAL 4 calculates the postplume passage doses based on a surface concentration immediately after plume passage that is inferred from the average concentration. In calculating the postplume passage doses, the code assumes that surface concentration varies with time starting at the end of plume passage because of radioactive decay and daughter ingrowth. The doses also account for weathering. Sections 7.2.1.1 and 7.2.1.2 describe the details of the weathering function used in groundshine dose calculations and the resuspension function used in the inhalation dose calculations.

FMDose infers the surface contamination at the end of plume passage from the average air concentration as follows:

$$\omega_i(\Delta T) = C_{ai} \times v_d \times \Delta T, \tag{7-4}$$

where $\omega_i(\Delta T)$ is the ground concentration of radionuclide i at the end of plume passage ΔT (ci/m²).

Using this value as the starting concentration, the FMDose calculates the portion of the early-phase groundshine dose occurring after plume passage using:

$$D_{gsio}(96 - \Delta T) = \omega_i(\Delta T) \times SRF \times DCF_{gsio} \int_0^{96-\Delta T} FC_{gi}(t) \times W_{gs}(t)dt, \tag{7-5}$$

where:

> $D_{gsio}(96\text{-}\Delta T)$ = the groundshine dose to organ o from radionuclide i during the portion of the early phase that follows plume passage (rem)
> SRF = a dimensionless surface roughness factor of 0.82 (Section 7.2.1.1)
> t = the time in hours following plume passage
> $FC_{gi}(t)$ = the fraction of radionuclide i remaining at time t following plume passage (dimensionless)
> $W_{gs}(t)$ = a dimensionless weathering function (Section 7.2.1.1)

FMDose calculates the inhalation doses for the postplume passage portion of the early phases using:

$$D_{inhio}(96 - \Delta T) = \omega_i(\Delta T) \times B \times DCF_{inhio} \int_0^{96-\Delta T} FC_{gi}(t) \times R(t)dt, \tag{7-6}$$

where $R(t)$ is the resuspension factor at time t (m⁻¹). The initial value of $R(t)$ is 1.0×10^{-5} m⁻¹.

Similarly, the FMDose calculates air submersion doses using:

$$D_{asio}(96 - \Delta T) = \omega_i(\Delta T) \times DCF_{io} \int_0^{96-\Delta T} FC_{gi}(t) \times R(t)dt \tag{7-7}$$

Note that the postplume passage portion of the early phase includes doses from daughters; they are not included in the plume passage portion of the early-phase dose calculation unless the measured concentrations include radionuclides with implicit daughters (Appendix A).

7.1.2 Ground Concentration Measurements

Early-phase dose calculations based on ground measurements assume that the measurements are made immediately following plume passage. These doses include groundshine doses, air submersion doses, and air inhalation doses.

7.1.2.1 Plume Passage Doses

During plume passage, the air submersion and inhalation doses assume that the individual is in the plume. Assuming that the plume is present at the measurement point at all times during plume passage, the FMDose estimates the groundshine dose to organ o from radionuclide i during plume passage as:

$$D_{gsio}(\Delta T) = \omega_i(\Delta T) \times \frac{\Delta T}{2} \times DCF_{gsio}, \qquad (7\text{-}8)$$

where $\omega_i(\Delta T)$ is the measured surface concentration of radionuclide i.

The air exposure (time integral of concentration) to radionuclide i during plume passage is inferred from the surface measurement using:

$$\int_0^{\Delta T} C_{ai} \times \Delta T \, dt = \frac{\omega_i(\Delta T)}{v_d} \qquad (7\text{-}9)$$

Thus, the air submersion dose to organ o from radionuclide i during plume passage is:

$$D_{asio}(\Delta T) = \frac{\omega_i(\Delta T)}{v_d} \times DCF_{asio}, \qquad (7\text{-}10)$$

the air inhalation committed dose to organ o from radionuclide i during plume passage is:

$$D_{ainhio}(\Delta T) = B \times \frac{\omega_i(\Delta T)}{v_d} \times DCF_{ainhio} \qquad (7\text{-}11)$$

Doses calculated during plume passage do not account for decay or daughter ingrowth, except for those radionuclides having implicit daughters. (Appendix A discusses the treatment of implicit daughters.)

7.1.2.2 Postplume Doses

Equations 7-5, 7-6, and 7-7 are used to calculate doses for both the portion of the early phase during plume passage and the portion of the early phase after plume passage. After plume passage, the doses are calculated from ground concentration measurements; during plume passage they are calculated using estimated concentrations. Doses calculated from ground concentration measurements include explicit treatment of decay and progeny ingrowth and decay because post plume passage exposure periods are longer than plume passage. However, the early-phase dose calculations do not include weathering. If these calculations include weathering, this parameter would change the doses by less than 0.3 percent.. Air submersion and inhalation doses are frequently small compared to the groundshine dose.

7.1.3 Total Effective and Absorbed Adjusted Bone Marrow Doses

RASCAL 4 calculates two composite doses. These are the total effective dose equivalent (TEDE) and the absorbed adjusted bone marrow dose (hereinafter referred to as the bone marrow dose). The TEDE is the sum of the groundshine dose, the submersion dose, and the effective inhalation dose. Groundshine and submersion dose conversion factors for TEDE doses are from FGR-12 (Eckerman and Ryman, 1993) as updated in DCFPAK 2 (Eckerman et al., 2008). The effective inhalation dose conversion factors are from either FGR-11 (Eckerman et al, 1988) as updated in DCFPAK2 or from ICRP-60 (ICRP, 1991). The cloud submersion component is likely to be zero (below 1×10^{-6} rem) when TEDE is calculated from ground concentrations. The bone marrow dose is the sum of the groundshine dose, the cloud submersion dose, and an inhalation dose to the bone marrow. The groundshine and submersion dose conversion factors used in calculating the bone marrow dose are from FGR-12 as updated, and an acute dose conversion factor from IAEA (2006) is used in the calculation of the inhalation dose to the bone marrow.

7.2 Intermediate-Phase Dose Calculations

RASCAL 4 calculates intermediate-phase doses from external exposure to, and inhalation of, radionuclides initially deposited on the ground. The intermediate phase of a radiological emergency begins after the release has terminated. Dose calculations for the intermediate phase are done to determine whether the concentrations of radionuclides on the ground are likely to cause doses to residents that would be in excess of the intermediate-phase protective action guides (PAGs) established by the U.S. Environmental Protection Agency (EPA) (EPA, 1992). The PAG is 2 rem for the first year. In addition, EPA specified an objective of 0.5 rem for the second year and 5 rem for the entire 50 years following the event.

In the "Field Measurements to Dose" module of RASCAL 4, the user enters the ground concentrations of deposited radionuclides at a location. Then, RASCAL 4 calculates the intermediate-phase doses for the first and second years and the cumulative dose over 50 years. The code calculates the doses with and without a delay in returning to the affected area. It calculates doses for selected delays ranging from 0 days to 40 years. The intermediate-phase doses calculated with delays in returning are for the remainder of the period. For example, if the return delay is 10 days, the first-year doses are for the period from day 11 through day 365. The second-year intermediate-phase doses are not affected by a 10-day delay. First-year intermediate-phase doses are 0.0 when the return delay is 365days. Therefore, RASCAL 4 does not calculate first year doses when the return delay exceeds 365 days. Similarly, second-year intermediate-phase doses are 0.0 for a return delay of 730 days, and RASCAL 4 does not calculate them for delays exceeding 730 days.

Users may also enter a delay to refine the first-year intermediate-phase dose estimates. This delay can be any number of days from 0 to 365 days. The dose shown in the intermediate-phase dose table for a user-entered delay is for the portion of the first year that follows the delay. The first-year dose savings that results from the user-entered delay is the difference between the first-year intermediate-phase doses with and without delay.

RASCAL 4 also calculates derived response levels (DRLs). A DRL is a measurable quantity that indicates that the deposited activity could result in an intermediate-phase dose equal to one of the intermediate-phase PAGs. One type of DRL is the closed window (gamma) dose rate in mR/hour, which corresponds to doses equal to the first-, second-, and 50-year intermediate-phase PAGs. The other type of DRL is the ground concentration of a marker radionuclide equivalent to a PAG. RASCAL 4 calculates DRLs for delay times ranging from 0 days to 40 years.

The following assumptions are among the implicit assumptions in these calculations that users must consider when using the results of the calculations:

- The surface in the area is flat, and no structures or terrain features are present in the vicinity of the measurement.

- The surface roughness in the area is uniform.

- The surface does not change as a function of time (e.g., snow or growth of vegetation).

Any significant deviation from these assumptions could cause errors in the intermediate-phase doses or DRLs. In addition, note that DRLs are specific to the location where the field measurement is made. A DRL may be used in the immediate vicinity of the measurement location; however, it should not be used for another location unless some assurance exists that the initial surface contamination of the second location was the same as that for the location where the initial measurements were made.

7.2.1 Intermediate-Phase Doses

The intermediate-phase dose from radionuclides deposited on the ground is the sum of three dose components: (1) external dose from ground, (2) internal dose from the inhalation of resuspended particles, and (3) external dose from submersion in a cloud of resuspended particles.

If the RASCAL 4 user specifies a reentry delay, the code calculates intermediate-phase doses starting at the specified time of reentry. This calculation can be useful in making decisions on reentry into previously evacuated areas. The value of a protective action depends on the amount of dose that can be averted by undertaking the protective action. Doses that are not averted by undertaking the action should not influence the decision on undertaking a protective action.

7.2.1.1 Groundshine Doses

FMDose calculates the groundshine dose $D_{gs}(T)$ for occupancy time interval T by summing the groundshine dose contributions from each deposited radionuclide i. The groundshine dose from a single radionuclide i that decays to a stable nuclide is:

$$D_{gsi}(T) = C_{gi}(0) \times SRF \times DCF_{gsi} \int_{t\,start}^{t\,end} FC_{gi}(t)W_{gs}(t)dt, \qquad (7\text{-}12)$$

where:

$D_{gsi}(T)$ = the external dose equivalent from groundshine from radionuclide i over time interval T (rem)
T = the intermediate-phase time interval corresponding to the first-, second-, or 50-year PAG
$C_{gi}(0)$ = the initial ground-surface concentration of radionuclide i (at time 0) (Ci/m^2),
SRF = surface roughness factor to convert from a flat infinite plane to a real-world surface (it has a default value of 0.82) (dimensionless)
DCF_{gsi} = dose conversion factor that gives the effective dose rate for exposure to groundshine from radionuclide i deposited on a flat infinite-plane ground surface from Table III.3 of FGR-12 [rem / (Ci-s m^{-2})] ,
$FC_{gi}(t)$ = the fraction of the initial ground-surface concentration of radionuclide i present at time t (dimensionless)
$W_{gs}(t)$ = the groundshine weathering factor at time t (dimensionless)
$t\,start$ = the start time for time interval T

t end = the end time for time interval *T*

The integral in the equation has units of time (seconds) and can be thought of as the effective exposure time to radionuclide *i*. It can be solved in closed form, but the solution is not shown here.

In Equation 7-7, *t* = 0 is the time of deposition of the radionuclides. The time of deposition must be used because the weathering factor is anchored to the time of deposition. In some cases, deposition could occur over several hours. In that case, either the time of peak deposition or the end of the deposition period can be used as *t* = 0. In actuality, a difference of a few hours or even a few days has negligible effect on the dose estimate.

The dose conversion factor for exposure to contaminated ground surface DCF_{gsi} is the effective dose rate coefficient in FGR-12 (Eckerman and Ryman, 1993) as updated in DCFPAK2 (Eckerman et al., 2008).

The groundshine weathering function $W_{gs}(t)$ (Anspaugh et al., 2002), which is based in large part on Chernobyl data, is the sum of two exponential terms:

$$W_{gs}(t) = C_1 e^{-\alpha t} + C_2 e^{-\beta t}, \tag{7-13}$$

where *t* is the time after deposition in days.

The first term describes the weathering during the first few years after deposition, and the second term describes long-term weathering. The values of C_1 and C_2 are 0.4 and 0.5, respectively, and the values of α and β are 1.26×10^{-3} d^{-1}, and 3.8×10^{-5} d^{-1}, respectively. These correspond to half-lives of 1.5 years and 50 years. The resulting combination reduces the effective surface concentration by a factor of about 0.84 at the end of the first year, by a factor of about 0.74 at the end of the second year, and by a factor of 0.30 at the end of 50 years. The Federal Radiological Monitoring and Assessment Center (FRMAC) and EPA have also recently adopted this weathering function.

The start time for the interval *T* for comparison to the first-year PAG will be either 0 days or the reentry time if specified by the RASCAL 4 user. The end time will be 1 year after reentry.

When a deposited radionuclide *i* decays to one or more radioactive daughters instead of to a stable nuclide, the groundshine dose from the daughters must be added to the groundshine dose from the parent nuclide *i*. In this case, Equation 7-7 becomes:

$$D_{gsi}(T) = C_{gi}(0) \times SRF \times \sum_{n=0}^{N} \left[DCF_{gsn} \times \int_{t\,start}^{t\,end} FC_{gn}(t)\, W_{gs}(t) dt \right], \tag{7-14}$$

where:

the summation is over the parent and all radioactive daughters, and $FC_{gn}(t)$ is the fractional ground concentration of radionuclide *n* relative to $C_{gi}(0)$, which is the initial concentration of the parent radionuclide *i* (dimensionless) corrected for decay.

If a nuclide has a short-lived daughter, RASCAL 4 treats the parent-daughter combination as a single nuclide with the parent's half-life. The code assumes that the implicit daughter is in radiological equilibrium with the parent. The effective dose rate coefficient *DCF* for the combination of the parent and daughter is the sum of the parent's effective dose rate coefficient from FGR-12 plus the daughter's effective dose rate coefficient (corrected for branching ratio). Appendix A discusses the treatment of these combinations of parents plus short-lived daughters in detail. Table A-1 lists parent radionuclides with implicit daughters.

Groundshine data following the Fukushima reactor accident indicate that weathering during the first year following deposition may occur faster than predicted by Equations 7-13 and 7-14. The data indicate that the groundshine dose rate for long-lived nuclides at the end of the first year is reduced by a factor of about 0.6 (Kreek 2012). The difference in weathering between that observed at Fukushima and the weathering predicted by Equation 7-13 may be caused by differences in the type and amount of precipitation that occurred after the Fukushima and Chernobyl accidents.

RASCAL 4 will use the weathering function based on the Chernobyl accident until the differences are fully understood. However, users may adjust RASCAL intermediate-phase groundshine dose estimates based on Fukushima data using the curves in Figure 7-1. The adjustment factors in Figure 7-1 assume that the increased weathering is only because of a change in the first term of Equation 7-13. The value of α in Equation 7-13 that is implicit in the adjustment factor curves is 1.26×10^{-2} d^{-1}. Adjustment of the first- and 50-year groundshine doses should be made on a nuclide-by-nuclide basis because the adjustment factor is a function of half-life. Although the adjustment factor for the second-year groundshine dose is also a function of half-life, the variation in adjustment factor is small, and the contribution of short-lived nuclides to the total groundshine dose is negligible.

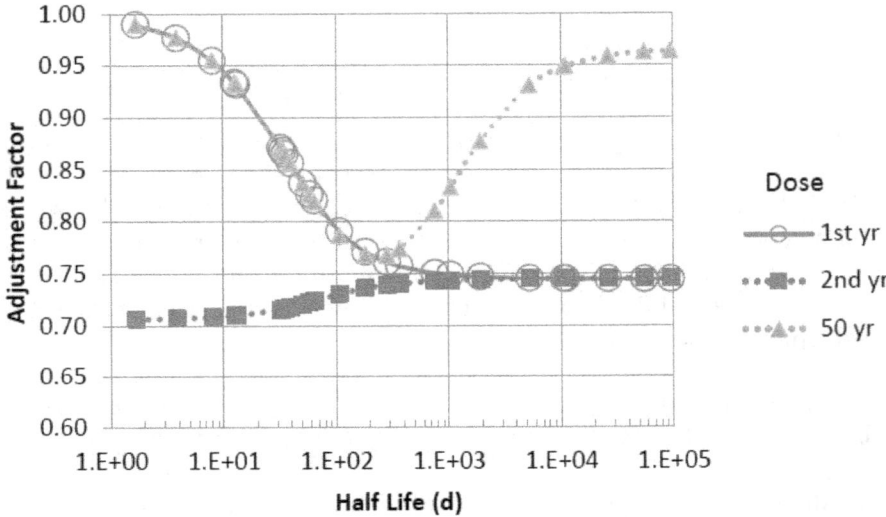

Figure 7-1   Intermediate-phase groundshine dose adjustment factors based on dose rate data collected following the Fukushima reactor accidents

7.2.1.2 Inhalation Doses

Intermediate-phase inhalation doses are generally much smaller than groundshine doses for nuclear power plant accidents. The doses are calculated assuming that activity on the ground is resuspended and then inhaled. The calculation, which is done separately for each radionuclide, is:

$$D_{inhi}(T) = C_{gi}(0) \times B \times DCF_{inh} \times \int_{t\,start}^{t\,end} FC_{gi}(t)\, R(t)\, dt, \qquad (7\text{-}15)$$

where:

$D_{inhi}(T)$ = the CEDE from inhalation of resuspended particles of radionuclide i over time interval T (rem)

$C_{gi}(0)$ = the initial ($t = 0$) ground-surface concentration of radionuclide i (Ci/m^2),

B = breathing rate (m³/s),
DCF_{inhi} = dose conversion factor for radionuclide i to calculate the CEDE from inhaled activity from Table 2.1 in FGR-11 (based on ICRP-30) or from ICRP-60 (ICRP, 1991) (rem-s/Ci)
$FC_{gi}(t)$ = the fraction of the initial ground-surface concentration of radionuclide i present at time t (dimensionless)
$R(t)$ = resuspension factor at time t (m⁻¹)
$t\ start$ = the start time for time interval T
$t\ end$ = the end time for time interval T

The default breathing rate for the intermediate-phase calculations B is 2.56×10^{-4} m³/s, which represents a long-term breathing rate that includes both waking and sleeping breathing rates (ICRP-66).

FMDose implements the Maxwell-Anspaugh (2011) resuspension factor model. In that model, the resuspension factor $R(t)$ is:

$$R(t) = A_1 e^{(B_1 t)} + A_2 e^{(B_2 t)} + C,\tag{7-15}$$

where:

$R(t)$ = resuspension factor (1/m)
A_1, A_2, C = constants with values of 1.93×10^{-6}, 1.71×10^{-8}, and 1.00×10^{-9}, respectively (1/m)
B_1, B_2 = constants with values of -0.039 and -0.0025, respectively (1/days)
t = time after deposition in days

The resuspension factor at time zero $R(0)$ is $A1 + A2 + C$, which has a value of 1.93×10^{-6} 1/m.

Figures 7-2 and 7-3 compare the Maxwell-Anspaugh model resuspension factors to the resuspension factors used in RASCAL prior to RASCAL 4.2. Although the Maxwell-Anspaugh resuspension factors are larger than those used in RASCAL prior to RASCAL 4.2 for the first few years after deposition, the increase in resuspension factors does not significantly increase intermediate-phase TEDE doses for most radionuclides because the portion of TEDE caused by inhaled radionuclides is generally several orders of magnitude smaller than the portion caused by groundshine and submersion. However, for a few radionuclides, such as strontium-90, the increase in TEDE is significant.

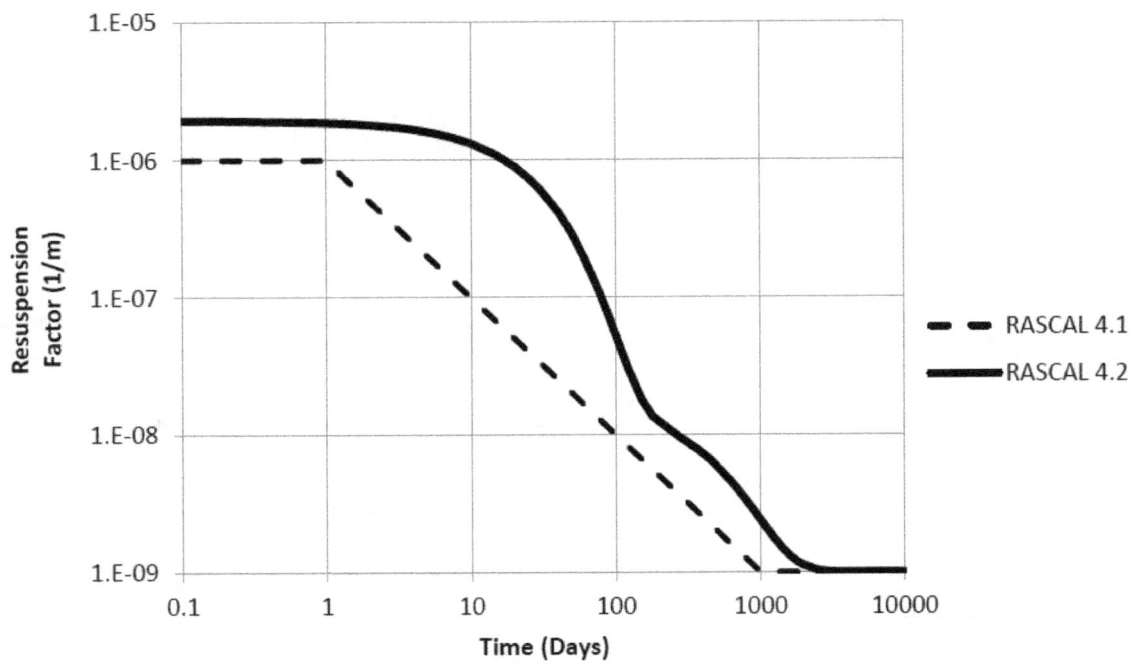

Figure 7-2 Comparison of resuspension-factor time dependence for resuspension factors used in RASCAL 4.2 and those used in earlier versions of RASCAL

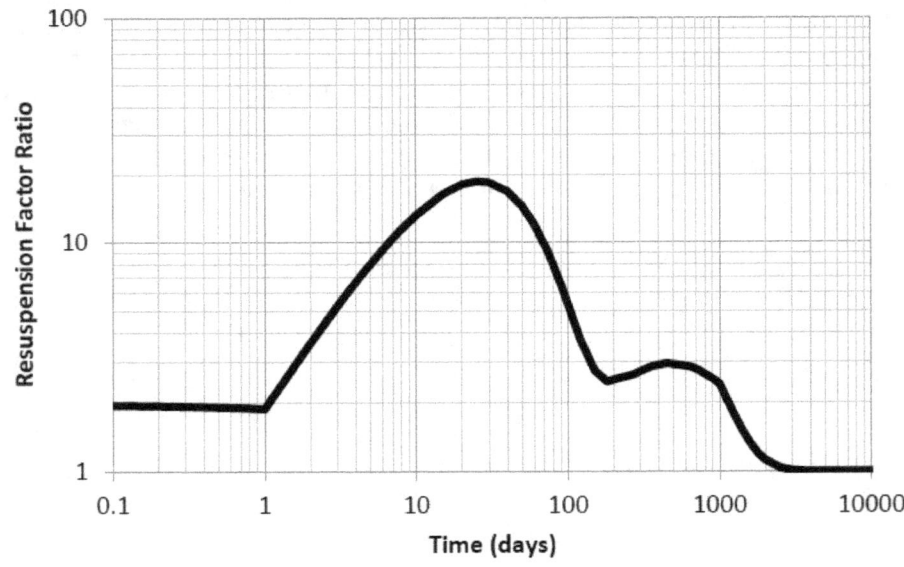

Figure 7-3   Increase in RASCAL 4.2 resuspension factors over those used in prior versions of RASCAL

When a deposited radionuclide i decays to one or more radioactive daughters instead of to a stable nuclide, the inhalation dose from the daughters must be added to the dose from the parent nuclide i. The summation is the same as that done for groundshine dose in Equation 7-14. It is:

$$D_{inhi}(T) = C_{gi}(0) \times B \times \sum_{n=0}^{N} \left[DCF_{inhn} \times \int_{t\,start}^{t\,end} FC_{gn}(t)\, R(t)dt \right] \tag{7-17}$$

7.2.1.3 Submersion Doses

Intermediate-phase submersion doses are generally smaller than either groundshine or inhalation doses. However, RASCAL 4.2 includes them for completeness. The calculations of the doses are done under the assumption that activity on the ground is resuspended and that an individual is submersed in a cloud of resuspended activity. FMDose calculates this submersion dose separately for each radionuclide using:

$$D_{asi}(T) = C_{gi}(0) \times DCF_{asi} \times \int_{t\,start}^{t\,end} FC_{gi}(t)\, R(t)dt, \tag{7-18}$$

where:

$D_{asi}(T)$ = the submersion dose for radionuclide i over time interval T (rem)
$C_{gi}(0)$ = the initial ($t = 0$) ground surface concentration of radionuclide i (Ci/m^2),
DCF_{asi} = submersion dose conversion factor for radionuclide i from Table 3.1 in FGR-11 (based on ICRP-30) as updated ((rem/s)/(Ci/m^3))
$FC_{gi}(t)$ = the fraction of the initial ground-surface concentration of radionuclide i present at time t (dimensionless)
$R(t)$ = resuspension factor at time t (m^{-1})
$t\,start$ = the start time for time interval T
$t\,end$ = the end time for time interval T

When a deposited radionuclide i decays to one or more radioactive daughters instead of to a stable nuclide, the submersion dose from the daughters must be added to the dose from the parent nuclide i. The summation is the same as that done for groundshine dose in Equation 7-14 and for inhalation dose in Equation 7-17. It is:

$$D_{asi}(T) = C_{gi}(0) \times \sum_{n=0}^{N} \left[DCF_{asin} \times \int_{t\,start}^{t\,end} FC_{gn}(t)\, R(t)dt \right] \tag{7-19}$$

7.2.1.4 Total Effective Dose Equivalents

When the user enters the ground concentrations of deposited radionuclides C_{gi}, RASCAL 4 calculates the TEDE for intermediate-phase occupancy time intervals T of first year, second year, and 50 years. If the user specifies a return delay, RASCAL 4 calculates the first-year dose with and without the delay.

The calculation of the intermediate-phase TEDE for the occupancy time interval is done by summing the external dose equivalent from groundshine, the internal CEDE from inhalation of resuspended particles, and the external dose from submersion in the cloud of resuspended particles. The summation is also over all radionuclides in the mixture entered by the user. The equation is:

$$TEDE(T) = \sum_i D_{gsi}(T) + \sum_i D_{inhi}(T) + \sum_i D_{asi}(T), \tag{7-20}$$

where:

> $TEDE(T)$ = the TEDE for intermediate-phase time interval T
> $D_{gsi}(T)$ = the external dose equivalent from groundshine over time interval T calculated from Equation 7-9 (rem)
> $D_{inhi}(T)$ = the CEDE over time interval T from inhalation of resuspended particles calculated from Equation 7-17 (rem)
> $D_{asi}(T)$ = the submersion dose over time interval T from submersion in the cloud of resuspended particles calculated from Equation 7-19 (rem)

RASCAL 4 output shows the groundshine dose, the inhalation dose, the submersion dose, and the sum of those doses for each intermediate-phase occupancy time interval T for the radionuclide mixture that the user entered. If the user specifies a delay time, the code calculates the first-year intermediate-phase dose and shows it for both the delay case and the no-delay case.

7.2.2 Dose Adjustments

Intermediate-phase doses calculated by RASCAL 4 may be adjusted for time in the contaminated area, building occupancy, and building shielding. The adjustments are made using information entered on the calculation options. As a default, RASCAL 4 assumes that a person is in the contaminated area 168 hours per week and does not enter buildings. The dose for an individual working in the area 40 hours per week and spending half of the time in a building would be significantly lower than the dose calculated using the default RASCAL assumptions. The dose adjustment factor is:

$$AF = \frac{AO}{168} \times [BO \times BS + (1 - BO)], \tag{7-21}$$

where:

> AF = the adjustment factor (dimensionless)
> AO = the number of hours per week in the contaminated area
> 168 = the number of hours in a week
> BO = the fraction of time in buildings in the contaminated area (dimensionless)
> BS = the building shielding factor (dimensionless)

Guidance on building occupancy fractions (BO) and building shielding factors (BS) appears in Chapter 3 of National Council on Radiation and Measurements (NCRP) Report No. 129, "Recommended Screening Limits for Contaminated Surface Soil and Review of Factors Relevant to Site-Specific Studies," (NRCP, 1999), and Tables 6.1 and 6.2 of the FRMAC manual (FRMAC, 2010) provide guidance related to building shielding fractions. Building occupancy fractions range from 0 to about 0.9, depending on the type of area. Values near zero are appropriate for agricultural areas, whereas values between 0.2 and 0.9 are appropriate for suburban areas. Shielding factors range from near zero to 0.5, with typical values in the 0.2 to 0.4 range.

7.2.3 Derived Response Levels

A DRL is a measurable quantity that indicates that the deposited activity could result in an intermediate-phase dose equal to one of the intermediate-phase PAGs. One type of DRL is the closed window (gamma) dose rate in mR/hour which is equal to the first-, second-, or 50-year intermediate-phase PAG. The other DRL is the ground concentration of a marker radionuclide equivalent to a PAG.

RASCAL 4 computes both sets of DRLs—one set for use with measured exposure rates (meter readings) and the other set for use with measurement of the surface contamination of a marker radionuclide. In each case, the DRLs are based on an assumed mixture of radionuclides on the surface. Consequently, DRLs are strictly applicable only to the location where the initial measurements were made. As a practical matter, they may be used with caution near the measurement site. However, they should not be used at locations where the deposited nuclide mix could be significantly different than the measured mix.

RASCAL 4 computes DRLs for the first and second years and for 50 years for a range of measurement times and return times from 0 to 40 years after the initial measurement of the radionuclide mix. The DRLs calculated by RASCAL assume a 168-hour-per-week occupancy and with no time in buildings. When RASCAL 4 calculates DRLs for periods with return delays, the DRLs are for the remainder of the period. For example, the first-year DRL for a delay of 30 days in returning to the affected area is the DRL that corresponds to the dose accrued in days 31–365 that would equal the first-year PAG.

7.2.3.1 Exposure Rate Derived Response Levels

The exposure rate DRL DRL_{exp} is the exposure rate that corresponds to a ground concentration that would result in a dose to an inhabitant equal to the intermediate-phase PAG. Thus, an exposure rate measurement survey instrument can be used to identify areas where doses might exceed the intermediate-phase PAGs.

The radionuclide mix deposited from a nuclear power plant accident is likely to contain many short-lived radionuclides and some longer lived radionuclides. During the first week or two, the exposure rate is likely to drop rapidly with time because the short-lived radionuclides contribute most of the groundshine dose during the early weeks. In fact, the exposure rate 1 day after deposition is likely to be significantly lower than the exposure rate at the time of deposition. Therefore, noting the measurement time for each reading and using a DRL_{exp} calculated for that measurement time are important.

In addition, using a DRL_{exp} calculated for the intended return time is important. It will generally take a week or two to have enough measurements and analyses to accurately define the areas that exceed an intermediate-phase PAG. PAGs are forward-looking; they include only future doses that can be averted by undertaking an action, such as long-term relocation. Thus, if the "return into an area" were being evaluated as taking place 1 week after an accident, the avertable first-year intermediate-phase dose would be the dose starting 1 week after the accident and ending 1 year after the accident.

The equations below define time $t = 0$ as the time of deposition; all other times are measured from that time. It is essential to define $t = 0$ as the time of deposition because the time-dependent functions for the groundshine weathering factor and the resuspension factor are anchored to the time of deposition.

The following ratio defines the exposure rate DRL for measurement time t_m and occupancy time internal T, $DRL_{exp}(t_m, T)$:

$$\frac{DRL_{exp}(t_m, T)}{PAG(T)} = \frac{DR(t_m)}{KD \times TEDE(T)},$$

(7-22)

where:

$DRL_{exp}(t_m, T)$ = exposure rate DRL for an exposure rate measurement made at time t_m for the occupancy time interval T (mR/hr),
$PAG(T)$ = EPA PAG dose for occupancy time interval T (rem)
$DR(t_m)$ = the dose rate from all radionuclides (including ingrowth of daughters) at measurement

time t_m (rem/hr),
KD = conversion factor from dose to kerma (0.7 (EPA, 1992, page 7-11)) (rem/R)
(Cember (1983) discusses kerma.)
$TEDE(T)$ = the TEDE from all deposited radionuclides i (including contributions from daughters) over time interval T as defined in Equation 7-12

FMDose uses the following basic equation for exposure rate DRLs:

$$DRL_{exp}(t_m, T) = PAG \times \frac{SRF \times \sum_n C_{gn}(t_m) \times DCF_{gsn}}{KD \times \sum_i [D_{gsi}(T) + D_{inhi}(T)]} \times 3600, \qquad (7\text{-}23)$$

where:

$C_{gn}(t_m)$ = the ground concentration of radionuclide n at measurement time t_m (Ci/m^2)
SRF = surface roughness factor with a default value of 0.82 (dimensionless)
DCF_{gsn} = dose conversion factor that gives the effective dose rate for exposure to groundshine from radionuclide n on a flat infinite-plane ground surface from Table III.3 of FGR-12 (rem/(Ci s m^{-2}))

7.2.3.2 Marker Nuclide Derived Response Levels

In some instances it may be difficult to use the exposure rate DRL to identify areas where doses might exceed the PAGs. Examples are when the exposure rate is near background levels or when there are no gamma-emitting radionuclides in the mix. In those instances, measuring the surface concentration of a particular marker radionuclide instead of the exposure rate may be easier.

The marker nuclide DRL is the concentration of that nuclide that would cause doses to inhabitants equal to the PAGs. The following ratio defines the marker DRL for measurement time t_m and occupancy time internal T, $DRL_{mark}(t_m, T)$:

$$\frac{DRL_{mark}(t_m, T)}{PAG(T)} = \frac{C_{mark}(t_m)}{TEDE(T)}, \qquad (7\text{-}24)$$

where:

$DRL_{mark}(t_m, T)$ = marker nuclide DRL at time t_m for the occupancy time interval T (Ci/m^2),
$PAG(T)$ = EPA PAG dose for occupancy time interval T (rem)
$C_{mark}(t_m)$ = the concentration of the marker nuclide at measurement time t_m (Ci/m^2).

7.3 Code Verification

The code used for calculating early- and intermediate-phase doses and DRLs in RASCAL 4 has undergone numerous tests. The tests demonstrate that the code does not contain any major errors. Code tests have included hand calculation checks of intermediate and final doses and DRL estimates, comparisons to doses and DRLs calculated using the manual methods described in the FRMAC manual (SNL, 2011), and comparisons of RASCAL 4 doses and DRLs to doses and DRLs from the TurboFRMAC 2011 code (TF 2011) (SNL, 2009).

The comparisons presented in the following sections show dose estimates calculated by RASCAL 4.2 and TF 2011 for 30 radionuclides that are commonly included in nuclear reactor accident source terms. RASCAL 4.2 and TF 2011 use the same weathering and decay functions. Both codes used ICRP-60 inhalation dose conversion factors to calculate CEDEs. The updated FGR-12 groundshine and submersion dose conversion factors were used in calculating external doses.

7.3.1 Early-Phase Doses

Figures 7-4–7-6 compare early-phase doses inferred by the two codes from field measurements. The TF 2011 calculations were made using the TF 2011 four pathway model with the custom decay chain option and RASCAL 4.2 deposition velocities.

Figure 7-4 shows the inferred inhalation CEDEs. The five points falling below the line are for isotopes of iodine. The lower values estimated by RASCAL 4.2 are the result of the RASCAL's representation of iodines as a mixture of particles and gases.

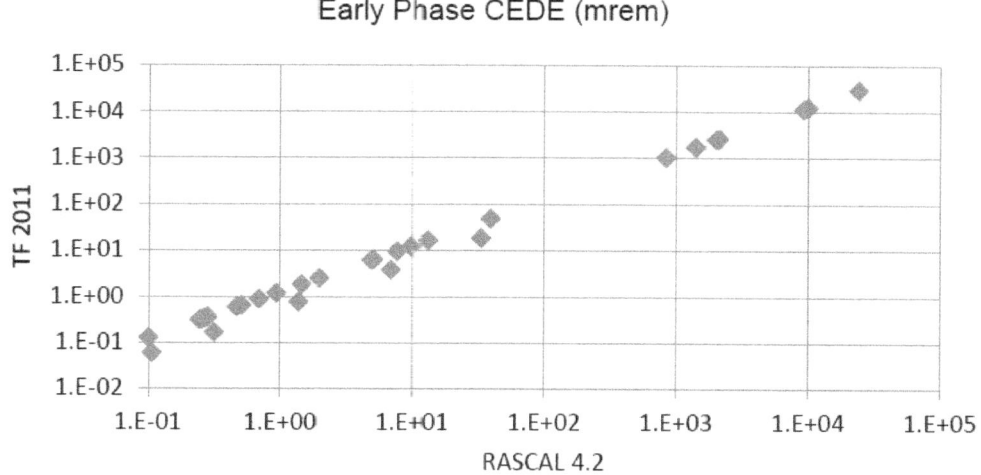

Figure 7-4   Comparison of RASCAL 4.2 and TF 2011 CEDEs estimated from field measurements

Figure 7-5 compares the external doses calculated for the same nuclides. The external dose includes doses from both cloudshine and groundshine. The iodines do not stand out in the external dose comparison because the dose rates for the iodines are not a function of the iodine form. The only significant deviation apparent in Figure 7-5 is for uranium-238. TF 2011 includes more daughters in the uranium-238 chain than are included in the simplified RASCAL 4.2 decay chain.

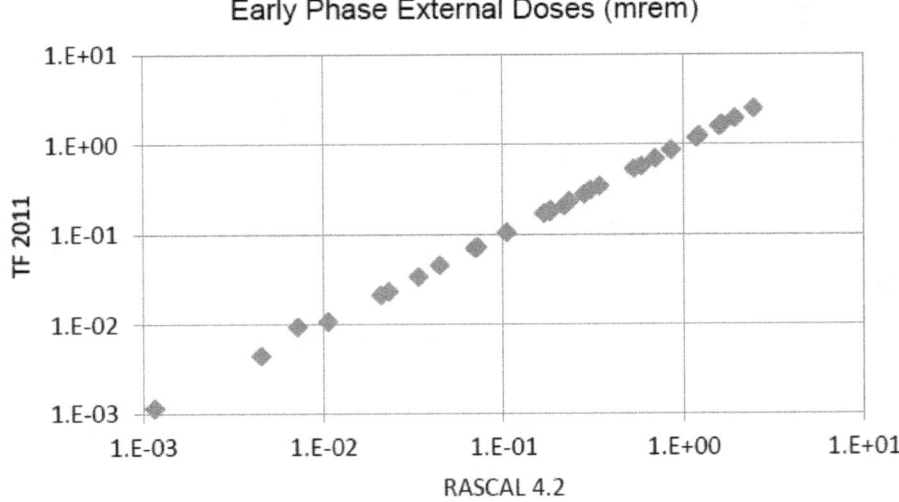

**Figure 7-5   Comparison of RASCAL 4.2 and TF 2011 external doses
estimated from field measurements**

Finally, Figure 7-6 compares early-phase TEDEs computed by the two codes. Again, the points deviating
from the line are for isotopes of iodine.

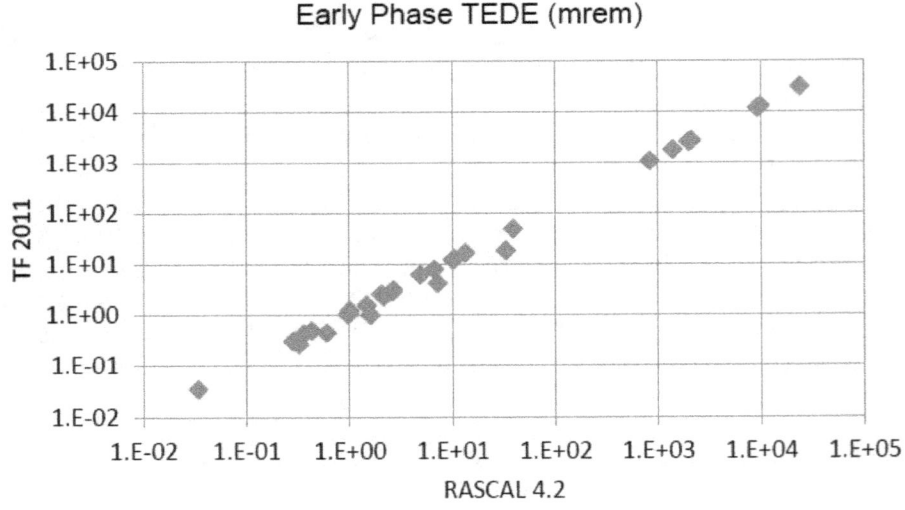

**Figure 7-6   Comparison of RASCAL 4.2 and TF 2011 TEDEs estimated
from field measurements**

7.3.2 Intermediate-Phase Doses

Figure 7-7 compares the first-, second-, and 50-year intermediate-phase TEDEs calculated by
RASCAL 4.2 and TF 2011. The TF 2011 intermediate-phase dose calculations were made using the
TF 2011 two pathway model and the custom decay chain option. The TF 2011 two pathway model was
used to avoid inclusion of the early-phase inhalation and groundshine doses in the intermediate-phase
doses.

The differences between the RASCAL 4.2 and TF 2011calculations are generally sufficiently small that they are insignificant for all practical purposes. In contrast with the early-phase CEDE and TEDE doses, the RASCAL 4.2 first- and 50-year intermediate-phase doses for the iodine isotopes do not differ significantly from those calculated by TF 2011. RASCAL 4.2 does not report doses for iodines for the second year because the half-lives of the iodine isotopes (iodine-131 (I-131) through I-135) are only a small fraction of a year. While TF 2011 reports doses for these isotopes, the doses reported are extremely small.

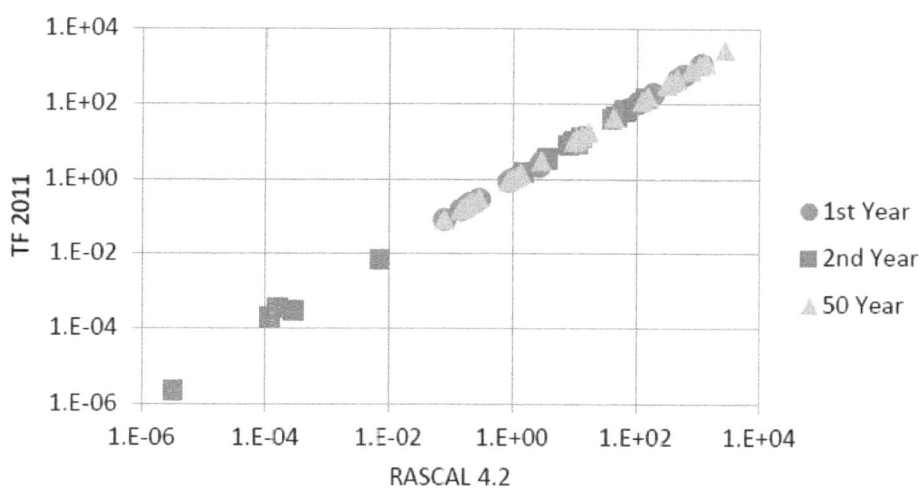

Intermediate Phase TEDE (mRem)

Figure 7-7   Comparison of RASCAL 4.2 and TF 2011 intermediate-phase TEDEs estimated from field measurements

Comparisons between RASCAL 4.2 and TF 2011 estimates of intermediate-phase doses from the inhalation of and the submersion in resuspended activity are similar to the comparisons shown in Figures 7-3 and 7-4. For most nuclides, the intermediate-phase inhalation and submersion doses are small fractions of the intermediate-phase groundshine doses. Notable exceptions to this generalization include I-131 (inhalation dose is about 8 percent of the groundshine dose) and isotopes of uranium and americium (inhalation doses exceed the groundshine doses).

7.3.3 Derived Response Levels

RASCAL 4.2 and TF 2011 calculate two types of DRLs. This section presents comparisons of the RASCAL 4.2 and TF 2011 estimates of each type of DRL.

The first type of DRL is based on surface contamination of a specified nuclide chosen to act as a surrogate for a mixture of nuclides. The DRL for the surrogate, called the marker nuclide, is the surface activity ($\mu Ci/m^2$) that corresponds to a specific EPA PAG. Figure 7-8 compares the marker DRL estimates of the two codes. The DRL estimates that the two codes use for comparison to the EPA first- and 50-year PAGs are nearly identical. The differences can generally be attributed to round-off errors in the code input and presentation of results.

Larger differences appear in the estimates of the DRL for comparison to the EPA second-year PAGs. The differences are generally found in DRLs for which the marker is a short-lived nuclide. For short-lived radionuclides, the second-year DRL calculation involves the ratio of two numbers, both of which are approaching zero. As a result, the differences between RASCAL 4.2 and TF 2011 DRLs tend to be larger than those for the first- and 50-year DRLs. Users should view DRLs for the second year with skepticism for radionuclides with half-lives of 15 days or less. RASCAL 4.2 includes a filter to avoid meaningless second-year DRLs.

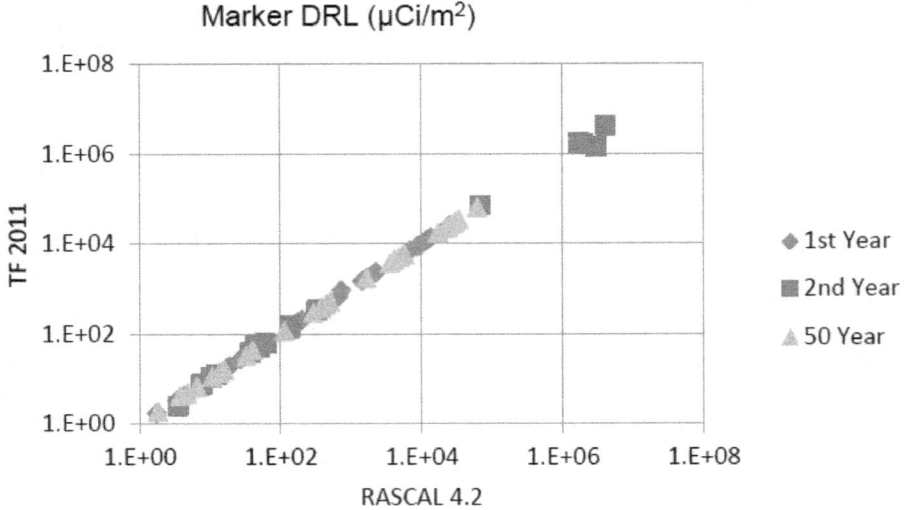

Figure 7-8 Comparison of RASCAL 4.2 and TF 2011 marker DRLs

The second type of DRL calculated by the two codes is the exposure rate DRL. These DRLs are an estimate of the exposure rates (mRad/hr) at the time of entry into a contaminated area that correspond to EPA PAGs. They are based on an assumed mixture of nuclides on the ground. Figure 7-9 compares exposure rate DRLs that are calculated by the two codes. Again, the DRLs that are calculated by the two codes are generally not significantly different. However, the comparison does show that one nuclide, cesium-137, has significantly different exposure rate DRLs. RASCAL 4.2 includes the exposure to the short-lived cesium-137 daughter, barium-137m, in the calculation of the DRL; TF 2011 does not. A careful examination of Figure 7-9 also shows that the DRLs have a small systematic difference. That difference is caused by differences in the conversion from dose rate (mrem/hr) to exposure rate (mRad/hr). RASCAL 4.2 uses a conversion of 1.0 mrem = 0.7 mR. TF 2011 uses a conversion of 1 mrem = 1 mR.

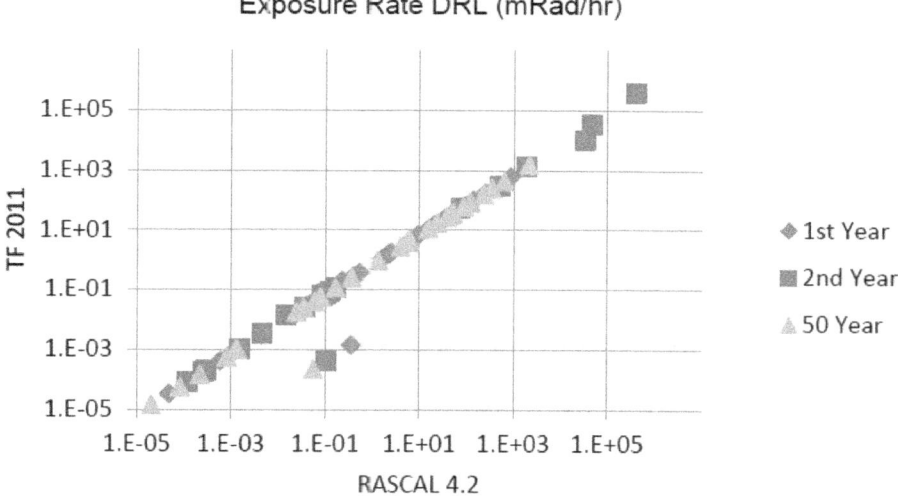

Figure 7-9   Comparison of RASCAL 4.2 and TF 2011 exposure rate DRLs

7.4 References

Alpert, D.J., D.I. Chanin, and L.T. Ritchie. 1986. "Relative Importance of Individual Elements to Reactor Accident Consequences Assuming Equal Release Fractions," NUREG/CR-4467, U.S. Nuclear Regulatory Commission, Washington, DC.

Anspaugh, L.R., S.L. Simon, K.I. Gordeev, I.A. Likhtarev, R.M. Maxwell, and S.M. Shinkarev. 2002. "Movement of Radionuclides in Terrestrial Ecosystems by Physical Processes," *Health Physics*, 82(5):669–679.

Benedict, M., T.H. Pigford, and H.W. Levi. 1987. *Nuclear Chemical Engineering*, Second Edition, McGraw-Hill, New York, NY.

Cember, H. 1983. *Introduction to Health Physics*, Pergamon Press, New York, NY.

Chanin, D.I., and M.L. Young. 1998. "Code Manual for MACCS2: Volume 1, 'User's Guide,'" NUREG/CR-6613, U.S. Nuclear Regulatory Commission, Washington, DC.

Eckerman, K.F., A.B. Wobarst, and A.C.B. Richardson. 1988. "Limiting Values of Radionuclide Intake and Air Concentration and Dose Conversion Factors for Inhalation, Submersion, and Ingestion," Federal Guidance Report No. 11, EPA-520/1-88-020, U.S. Environmental Protection Agency, Washington, DC.

Eckerman, K.F., and J.C. Ryman. 1993. "External Exposure to Radionuclides in Air, Water, and Soil," Federal Guidance Report No. 12, EPA-402-R-93-081, U.S. Environmental Protection Agency, Washington, DC.

International Commission on Radiological Protection (ICRP). 1991. "1990 Recommendations of the International Commission on Radiological Protection," ICRP Publication 60, *Annals of the ICRP*, 21:1–3.

International Atomic Energy Agency (IAEA). 2006. *Dangerous Quantities of Radioactive Materials, EPR-D-Values*, Vienna, Austria.

Kreek, S. 2012. "Description of Observations Collected around the Fukushima Site by the U.S. Department of Energy Teams." Presented at International Workshop on Source Term Estimation (STE) Methods for Estimating the Atmospheric Radiation Release from the Fukushima Daiichi Nuclear Plant, National Center for Atmospheric Research, Boulder, CO.

Maxwell, R.M., and L.R. Anspaugh. 2011. "An Improved Model for Prediction of Resuspension," *Health Physics,* 101:6.

National Council on Radiation Protection and Measurements (NCRP). 1999. "Recommended Screening Limits for Contaminated Surface Soil and Review of Factors Relevant to Site-Specific Studies," NCRP Report No. 129, National Council on Radiation Protection and Measurements, Bethesda, MD.

Sandia National Laboratories (SNL). 2010. "FRMAC Assessment Manual: Volume 1, 'Overview and Methods,'" SAND2010-1405P, Sandia National Laboratories, Albuquerque, NM, 2011.

Sandia National Laboratories (SNL). 2009. "TurboFRMAC 2009," Sandia National Laboratories, Albuquerque, NM.

Strenge, D.L. 1997. "A General Algorithm for Radioactive Decay with Branching and Loss from a Medium," *Health Physics,* 73:953–957.

U.S. Environmental Protection Agency (EPA). 1992. "Manual of Protective Action Guides and Protective Actions for Nuclear Incidents," EPA-400-R-92-001, U.S. Environmental Protection Agency, Washington, DC.

APPENDIX A
RADIOLOGICAL DECAY AND INGROWTH

The dose calculations in RASCAL 4 involve the calculation of the decay of radionuclides and the ingrowth and decay of progeny. The following discussion presents the equations used to calculate decay and ingrowth. It also discusses modifications made in decay chains to simplify the RASCAL 4 early- and intermediate-phase dose calculations. The simplifications include the combining of parents and short-lived progeny, the modification of chains that have multiple paths to a common daughter, and the truncation of decay chains when a long-lived daughter is encountered.

A.1 Radionuclide Library

The radionuclide library for RASCAL 4 includes most of the isotopes listed in Federal Guidance Report No. (FGR)-11, "Limiting Values of Radionuclide Intake and Air Concentrations and Dose Conversion Factors for Inhalation, Submersion, and Ingestion," (Eckerman et al., 1988), and FGR-12, "External Exposure to Radionuclides in Air, Water, and Soil," (Eckerman and Ryman, 1993). The isotopes listed in the FGRs that have been omitted from the RASCAL 4 library have half-lives of less than 10 minutes. These isotopes are omitted because the atmospheric transport and dispersion modules of RASCAL 4 use numerical procedures that do not adequately treat the decay and ingrowth of isotopes with half-lives of less than 10 minutes. Typically, these isotopes do not contribute significantly to doses.

RASCAL 4 recognizes 800 isotopes explicitly and three "special isotopes." The isotopes that RASCAL 4 recognizes include all 60 of the isotopes generally used for severe accident consequence assessments in the MACCS code (Alpert et al., 1986; Chanin and Young, 1998) and all but four of the isotopes listed in the FRMAC manual (SNL, 2010). The four isotopes listed in the FRMAC manual that are not explicitly included in the RASCAL 4 subset of isotopes include rhodium-106 (Rh-106) (half-life of 29.9 seconds), barium-137m (Ba-137m) (half-life of 2.52 minutes), and praseodymium-144m (Pr-144m) (half-life of 17.28 minutes). Each of these isotopes is included implicitly with its parent. The other FRMAC isotope that is not included in the RASCAL 4 subset is silver-109m (Ag-109m) (half-life of 39.6 seconds). The FRMAC manual includes Ag-109m with cadmium-109 (Cd-109). RASCAL 4 does not include Ag-109m because of its short half-life and because it is not in the Cd-109 decay chain in FGR-12. The "special isotopes" included in RASCAL 4 are "U-Enrch" (enriched uranium), "U-Natrl" (natural uranium), and uranium hexafluoride (UF$_6$). These "special isotopes" have sufficiently long half-lives so that decay is not significant, and they have no daughters.

In many cases, the short-lived isotopes that are not present explicitly in RASCAL 4 are included implicitly with the parent under the assumption that they exist in secular equilibrium with the parent. Table A-1 lists radionuclides that include implicit daughters and their implicit daughters. In RASCAL 4, parent radionuclides with implicit daughters are indicated by adding an * or a + symbol to the radionuclide name. For example cesium-137* (Cs-137*) is Cs-137 combined with its implicit daughter Ba-137m (half-life of 2.52 minutes), and strontium-90 (Sr-90+) includes its yttrium-90 (Y-90) daughter (half-life of 64 hours). In general, those radionuclides ending in the * symbol are intended for use in early-phase dose calculations, whereas those ending in the + symbol should only be used for intermediate-phase doses (e.g., first-year doses, second-year doses, and 50-year doses or in cases in which they are included as a daughter in a decay chain). In many cases, more than one option exists for a parent radionuclide. The RASCAL 4 library may list a parent radionuclide both with and without implicit daughters, or it may list a parent with two different sets of implicit daughters.

Table A-2 lists composite dose conversion factors (DCFs) for radionuclides with implicit daughters. The composite DCFs include branching, where appropriate. The composite acute bone, colon, and lung DCFs are based on International Atomic Energy Agency (IAEA) relative biological effectiveness (RBE) weighted DCFs (IAEA, 2006); the remaining composite DCFs are based on DCFs in FGR-11 (Eckerman et al., 1988) and FGR-12 (Eckerman and Ryman, 1993).

Table A-3 lists those short-lived radionuclides that are not included in RASCAL 4 explicitly. Comments in the list indicate those radionuclides that are included implicitly in decay chains. Ultimately, only 15 isotopes listed in FGR-11 and FGR-12 are not included in RASCAL either explicitly or implicitly.

A.2 Decay and Ingrowth Equations

The equations below describe the activity of the parent and daughters as a function of time (Strenge, 1997). Given the initial activity of a parent radionuclide at time zero, the equations give the activities of the radionuclide and daughters, if any, following the measurement. They are based on the Bateman equations (Bateman, 1910; Benedict et al., 1987), which deal specifically with atoms. Appendix B to NUREG/CR-5512, "Residual Radioactive Contamination from Decommissioning," Volume 1, "Technical Basis for Translating Contamination Levels to Annual Total Effective Dose Equivalent," (Kennedy and Strenge, 1992), contains the algebraic details of the modifications necessary to obtain the equations for activity (disintegrations/time) below from the Bateman equations for number of atoms.

The parent radionuclide activity is described by a simple exponential decay. It is:

$$A_p(t) = A_p(0)e^{-\lambda_p t}, \tag{A-1}$$

where:

$A_p(t)$ = activity of the parent at time t following the measurement
$A_p(0)$ = initial activity at time $t = 0$
λ_p = decay constant of the parent

Assuming that none of the first daughter is present initially, the activity of a first-generation daughter radionuclide from ingrowth from the parent is:

$$A_{d1}(t) = A_p(0)\frac{f_{d1}\lambda_{d1}}{\lambda_{d1} - \lambda_p}\left(e^{-\lambda_p t} - e^{-\lambda_{d1} t}\right), \tag{A-2}$$

where:

$A_{d1}(t)$ = activity of the first-generation daughter at time t
f_{d1} = fraction of disintegrations of the parent that yield the daughter
λ_{d1} = decay constant for the daughter

Assuming that none of either the first generation daughter or the second generation daughter is present initially, the activity of a second-generation daughter radionuclide is:

$$A_{d2}(t) = A_p(0) \times f_{d1}\lambda_{d1}f_{d2}\lambda_{d2}$$
$$\times \left[\frac{e^{-\lambda_p t}}{(\lambda_{d1} - \lambda_p)(\lambda_{d2} - \lambda_p)} + \frac{e^{-\lambda_{d1} t}}{(\lambda_p - \lambda_{d1})(\lambda_{d2} - \lambda_{d1})} + \frac{e^{-\lambda_{d2} t}}{(\lambda_p - \lambda_{d2})(\lambda_{d1} - \lambda_{d2})}\right], \tag{A-3}$$

where:

$A_{d2}(t)$ = activity of the second-generation daughter at time t following the measurement
f_{d2} = fraction of first-generation daughter disintegrations that yield the second-generation daughter

λ_{d2} = decay constant for the second-generation daughter

Finally, the following equation gives the activity of a third-generation daughter radionuclide under the assumption that no activity of the daughter or of the preceding-generation daughters is initially present:

$$A_{d3}(t) = A_p(0) \times f_{d1}\lambda_{d1}f_{d2}\lambda_{d2}f_{d3}\lambda_{d3}$$
$$\times \left[\frac{e^{-\lambda_p t}}{(\lambda_{d1} - \lambda_p)(\lambda_{d2} - \lambda_p)(\lambda_{d3} - \lambda_p)} + \frac{e^{-\lambda_{d1} t}}{(\lambda_{d1} - \lambda_p)(\lambda_{d2} - \lambda_{d1})(\lambda_{d3} - \lambda_{d1})} \right. \tag{A-4}$$
$$\left. + \frac{e^{-\lambda_{d2} t}}{(\lambda_{d2} - \lambda_p)(\lambda_{d2} - \lambda_{d1})(\lambda_{d3} - \lambda_{d2})} + \frac{e^{-\lambda_{d3} t}}{(\lambda_{d3} - \lambda_p)(\lambda_{d3} - \lambda_{d1})(\lambda_{d3} - \lambda_{d2})} \right],$$

where:

> $A_{d3}(t)$ = activity of the third-generation daughter at time t following the measurement
> f_{d3} = fraction of second-generation daughter disintegrations that yield the third-generation daughter
> λ_{d3} = decay constant for the third-generation daughter

Using the simplified decay chains described in the following sections, the longest decay chains have only third-generation daughters. Although there may be more than one second-generation daughter for a parent radionuclide, the simplified decay chains include, at most, a single second- or third-generation daughter.

A.3 Modified Decay Chains

The following paragraphs describe the methods used to simplify radioactive decay chains so that Equations A-1–A-4 are sufficient. RASCAL 4 treats radionuclide combinations of a long-lived parent and a short-lived daughter as a single radionuclide. In addition to the inclusion of short-lived daughters implicitly with parent radionuclides, other modifications have been made to the decay chains used in the RASCAL 4 early- and intermediate-phase dose calculations. These modifications include (1) the truncation of decay chains at the first very long-lived daughters in the chain (i.e., long relative to the 50-year intermediate-phase period), (2) the truncation at the first noble gas with a half-life that exceeds a few minutes, and (3) the dropping of radionuclides that contribute less than 1 percent of the total effective dose equivalent (TEDE).

Long half-life daughters are used to truncate decay chains under the assumption that the ingrown activity of the daughter will not reach sufficient concentrations to contribute significantly to dose relative to the parent. This is particularly true in cases in which the half-life of the daughter is long compared to the half-life of the parent. For example, in the decay of tellurium-129 (Te-129), iodine-129 (I-129) is considered to be stable. The exception to this rule is that neptunium-239 (Np-239) decay does include plutonium-239 (Pu-239) as a member of the chain because of the potential dose importance of decay. Most of the truncation in the RASCAL 4 decay chains occurs for high atomic number parents (radon and higher).

Noble gas daughters that have half-lives of more than a few minutes are used to truncate the decay chain because the assumption is that the noble gas will become airborne and will be carried away. For example, the dose calculations do not include xenon-131m (Xe-131m) with a 11.9-day half-life as a daughter of I-131. Similarly, the dose calculations include Xe-135m (half-life of 15 minutes) as a daughter of I-135, and the chain is truncated at that point because the next nuclide in the chain, Xe-135, is a noble gas with a relatively long half-life (9 hours).

The contribution to TEDE for the first and second years and 50 years after deposition was evaluated for each decay chain member to identify potential decay chain simplifications. Radionuclides that did not contribute 1 percent of the TEDE for any of these periods were dropped. The zirconium-95 (Zr-95) decay chain provides an example of the use of this criterion. Because the niobium-95n (Nb-95m) daughter of Zr-95 contributed less than 1 percent of the TEDE dose, it was dropped from the chain. More frequently, the criterion resulted in the truncation of chains. In other cases, a daughter contributed more than 1 percent of the second-year TEDE; however, the second-year TEDE was a small fraction (<0.1 percent) of the first- or 50-year TEDE. These daughters were also dropped from the chains. The dropping of daughters using TEDE-related criteria did not result in a change of TEDE of 1 percent in any case.

With these simplifications, no intermediate-phase decay chain includes more than three generations of explicit daughters. Table A-4 lists the decay chains for all 803 isotopes in the RASCAL 4 radionuclide library. The columns in the table are organized into four groups. The first group deals with the parent radionuclide. It starts with the parent name and deposition type. In the three deposition types, 0 indicates a nondepositing radionuclide (noble gas), 1 indicates a halogen (e.g., chlorine, bromine, and iodine), and 2 indicates other radionuclides that would be associated with particles. The last two items in the group are the half-life in days and the number of daughters in the simplified RASCAL 4 decay chain. The remaining groups deal with daughter radionuclides. They start with the daughter name and the position of the immediate parent in the chain. For example, if the immediate parent is the start of the chain, the position is 1; if it is the first daughter in the chain, the position is 2. The last two items are the half-life of the daughter and the branching fraction leading to the daughter. This group is repeated for each daughter.

A.4 Code Verification

Several subroutines have been prepared to make the decay and ingrowth calculations described in this appendix. Verification of these subroutines was done by comparing their results to hand calculations. Implementation of the subroutines in RASCAL 4 components has been checked by hand calculations and comparisons of RASCAL 4 computational results to similar computations made in previous versions of RASCAL and to the computational results of other codes (e.g., TurboFRMAC, 2009). Chapters 1, 4, and 7 include examples of these comparisons.

A.5 References

Alpert, D.J., D.I. Chanin, and L.T. Ritchie. 1986. "Relative Importance of Individual Elements to Reactor Accident Consequences Assuming Equal Release Fractions," NUREG/CR-4467, U.S. Nuclear Regulatory Commission, Washington, DC.

Bateman, H. 1910. "Solution of a System of Differential Equations Occurring in the Theory of Radioactive Decay Transformation, *Proceedings of the Cambridge Philosophical Society*, 15:423–427.

Benedict, M., T.H. Pigford, and H.W. Levi. 1987. *Nuclear Chemical Engineering*, Second Edition, McGraw-Hill, New York, NY.

Chanin, D.I., and M.L. Young. 1998. "Code Manual for MACCS2: Volume 1, 'User's Guide,'" NUREG/CR-6613, U.S. Nuclear Regulatory Commission, Washington, DC.

Eckerman K.F., A.B. Wobarst, and A.C.B. Richardson. 1988. "Limiting Values of Radionuclide Intake and Air Concentrations and Dose Conversion Factors for Inhalation, Submersion, and Ingestion," Federal Guidance Report No. 11, EPA-520/1-88-202, U.S. Environmental Protection Agency, Washington, DC.

Eckerman K.F., and J.C. Ryman. 1993. "External Exposure to Radionuclides in Air, Water, and Soil," Federal Guidance Report No. 12, EPA-402-R-93-081, U.S. Environmental Protection Agency, Washington, DC.

International Atomic Energy Agency (IAEA). 2006. "Dangerous Quantities of Radioactive Materials, EPR-D-Values," Vienna, Austria.

Kennedy, W.E., and D.L. Strenge. 1992. "Residual Radioactive Contamination from Decommissioning," Volume 1, "Technical Basis for Translating Contamination Levels to Annual Total Effective Dose Equivalent," NUREG/CR-5512, U.S. Nuclear Regulatory Commission, Washington, DC.

Sandia National Laboratories (SNL). 2010. "FRMAC Assessment Manual," SAND2010-1405P, Sandia National Laboratories, Albuquerque, NM.

Strenge, D.L. 1997. "A General Algorithm for Radioactive Decay with Branching and Loss from a Medium," *Health Physics,* 73(6):953–957.

Table A-1   Radionuclides with Implicit Daughters

PARENT	IMPLICIT DAUGHTER (branching fractions)
Mg-28*	Al-28 (1.0)
Ni-66*	Cu-66 (1.0)
Zn-62*	Cu-62 (1.0)
Zn-69m+	Zn-69 (0.9997)
Ge-68+	Ga-68 (1.0)
Br-83+	Kr-83m (1.0)
Sr-80*	Rb-80 (1.0)
Sr-82*	Rb-82 (1.0)
Sr-90+	Y-90 (1.0)
Sr-91+	Y-91m (0.578)
Zr-97*	Nb-97m (0.947)
Zr-97+	Nb-97m (0.947) , Nb-97 (0.053)
Mo-99+	Tc-99m (0.876)
Ru-103+	Rh-103m (0.99974)
Ru-106*	Rh-106 (1.0)
Pd-103+	Rh-103m (1.0)
Ag-108m*	Ag-108 (0.089)
Ag-110m*	Ag-110 (0.0133)
Cd-115+	In-115m (1.0)
In-114m*	In-114 (0.957)
In-119m*	In-119 (0.025)
Sn-113+	In-113m (1.0)
Sn-126+	Sb-126m (1.0)
Sb-124n*	Sb-124m (1.0)
Te-129m+	Te-129 (0.65)
Te-131m+	Te-131 (0.222)
Te-132+	I-132 (1.0)
Te-133m+	Te-133 (0.13)
Te-134+	I-134 (1.0)
I-135+	Xe-135m (0.154)
Xe-122*	I-122 (1.0)
Cs-137*	Ba-137m (0.947)
Ba-126*	Cs-126 (1.0)
Ba-128*	Cs-128 (1.0)
Ce-134*	La-134 (1.0)
Ce-144*	Pr-144m (0.0178)
Ce-144+	Pr-144m (0.0178), Pr-144 (1.0)
Nd-138*	Pr-138 (1.0)
Sm-142*	Pm-142 (1.0)
W-178*	Ta-178m (1.0)
Os-180*	Re-180 (1.0)
Pb-211*	Bi-211 (1.0), Po-211 (0.0027), Tl-207 (0.9973)

PARENT	IMPLICIT DAUGHTER (branching fractions)
Pb-212+	Bi-212 (1.0), Po-212 (0.6407), Tl-208 (0.3593)
Bi-210m*	Tl-206 (1.0)
Bi-212*	Po-212 (0.6407), Tl-208 (0.3593)
Bi-213*	Po-213 (0.9784), Tl-209 (0.0216)
Bi-214*	Po-214 (0.9998)
At-211*	Po-211 (0.5830)
Rn-222*	Po-218 (1.0)
Rn-222+	Po-218 (1.0), Pb-214 (1.0), Bi-214 (1.0), Po-214 (0.9998)
Fr-222*	Ra-222 (1.0), Rn-218 (1.0), Po-214(1.0)
Ra-223*	Rn-219 (1.0), Po-215 (1.0)
Ra-223+	Rn-219 (1.0), Po-215 (1.0), Pb-211 (1.0), Bi-211 (1.0), Po-211 (0.0028), Tl-207 (0.9972)
Ra-224*	Rn-220 (1.0), Po-216 (1.0)
Ra-224+	Rn-220 (1.0), Po-216 (1.0), Pb-212 (1.0), Bi-212 (1.0), Po-212 (0.6407), Tl-208 (0.3593)
Ra-228+	Ac-228 (1.0)
Ac-224*	Fr-220 (0.1), At-216 (1.0)
Ac-225*	Fr-221 (1.0), At-217 (1.0)
Ac-225+	Fr-221 (1.0), At-217 (1.0), Bi-213 (1.0), Po-213(0.9784), Tl-209 (0.0216), Pb-209 (1.0)
Ac-227*	Fr-223 (0.0138)
Th-226*	Ra-222 (1.0), Rn-218 (1.0), Po-214 (0.9998)
Th-227+	Ra-223 (1.0), Rn-219 (1.0), Po-215 (1.0), Pb-211 (1.0), Bi-211 (1.0), Po-211 (0.0027), Tl-207 (0.9973)
Th-228+	Ra-224 (1.0), Rn-220 (1.0), Po-216 (1.0)
Th-234*	Pa-234m (0.998)
Th-234+	Pa-234m (0.998), Pa-234 (0.002)
Pa-227*	Ac-223 (0.85), Fr-219 (0.85), At-215 (0.85), Bi-211 (0.85), Tl-207 (0.848), Po-211 (0.0024)
U-240*	Np-240m (1.0)
Pu-244+	U-240 (1.0), Np-240m (1.0)
Cm-247+	Pu-243 (1.0)
Es-254+	Bk-250 (1.0)

Table A-2 Composite DCFs for Radionuclides with Implicit Daughters

	FGR-11 AND FGR-12 DCFs						RBE 30-DAY ACUTE DCFs			ICRP-60* DCFs	
	Inhalation (Sv/Bq)	Thyroid (Sv/Bq)	Ground (Sv/(Bq s m^{-2}))	Immersion (Sv/(Bq s m^{-3}))	Skin—GS (Sv/(Bq s m^{-2}))	Skin—Air (Sv/(Bq s m^{-2}))	Colon (Gy-Eq/Bq)	Lung (Gy-Eq/Bq)	Red Marrow (Gy-Eq/Bq)	Inhalation (Sv/Bq)	Thyroid (Sv/Bq)
Mg-28*	1.33E-09	1.78E-10	2.97E-15	1.53E-13	1.67E-14	2.71E-13	2.40E-09	2.60E-09	5.50E-10	1.27E-09	1.38E-10
Ni-66*	2.25E-09	1.21E-10	2.15E-16	5.51E-15	1.19E-14	7.88E-14	0	0	0	1.80E-09	7.06E-11
Zn-62*	5.57E-10	2.94E-11	1.53E-15	6.54E-14	1.54E-14	1.69E-13	0	0	0	5.46E-10	6.22E-11
Zn-69m+	2.31E-10	1.35E-11	4.19E-16	1.86E-14	2.59E-15	4.25E-14	6.47E-10	4.24E-10	1.71E-11	2.95E-10	2.54E-11
Ge-68+	1.40E-08	6.94E-10	9.97E-16	4.29E-14	1.00E-14	1.01E-13	2.33E-09	2.61E-08	2.96E-10	3.21E-08	1.52E-09
Br-83+	2.41E-11	3.29E-12	2.91E-17	5.11E-16	2.18E-15	1.88E-14	0	0	0	5.26E-11	2.88E-12
Sr-80*	1.36E-10	1.30E-11	1.74E-15	7.46E-14	1.91E-14	2.37E-13	0	0	0	1.36E-10	1.12E-11
Sr-82*	1.66E-08	1.21E-09	1.21E-15	5.09E-14	1.54E-14	1.59E-13	1.00E-08	4.1E-08	4.30E-09	1.07E-08	5.97E-10
Sr-90+	3.53E-07	2.65E-09	1.12E-16	8.89E-16	1.06E-14	7.15E-14	8.39E-09	4.48E-08	3.70E-09	1.59E-07	6.28E-10
Sr-91+	4.55E-10	4.12E-11	1.03E-15	4.68E-14	8.11E-15	9.99E-14	4.20E-09	1.7E-08	5.31E-10	4.15E-10	2.69E-11
Zr-97*	1.17E-09	9.56E-11	9.18E-16	4.07E-14	9.41E-15	9.59E-14	2.90E-09	1.9E-09	3.05E-10	9.84E-10	7.08E-11
Zr-97+	1.19E-09	9.69E-11	1.60E-15	7.11E-14	1.50E-14	1.62E-13	2.92E-09	1.95E-09	3.52E-10	1.03E-09	7.36E-11
Mo-99+	1.08E-09	1.61E-10	2.77E-16	1.15E-14	3.90E-15	3.77E-14	2.01E-09	2.41E-09	2.01E-10	1.01E-09	1.19E-10
Ru-103+	2.42E-09	5.97E-10	4.75E-16	2.21E-14	6.20E-16	2.86E-14	2.80E-09	4.5E-09	4.10E-10	2.95E-09	3.35E-10
Ru-106*	1.29E-07	1.37E-08	3.46E-16	1.07E-14	1.42E-14	1.09E-13	2.70E-08	5.5E-08	2.60E-09	6.60E-08	7.13E-09
Pd-103+	4.25E-10	4.17E-12	8.49E-18	5.85E-17	6.67E-17	4.31E-16	3.31E-10	1.2E-09	8.39E-12	4.49E-10	2.16E-12
Ag-108m*	7.66E-08	2.01E-08	1.55E-15	7.24E-14	2.62E-15	9.37E-14	2.50E-09	7.1E-09	7.50E-10	3.88E-08	1.18E-08
Ag-110m*	2.17E-08	6.39E-09	2.59E-15	1.28E-13	3.45E-15	1.59E-13	2.50E-09	9.5E-09	1.20E-09	1.23E-08	3.73E-09
Cd-115+	1.18E-09	8.70E-11	3.59E-16	1.57E-14	2.75E-15	4.68E-14	2.19E-09	2.37E-09	8.90E-11	1.13E-09	5.46E-11
In-114m*	2.40E-08	2.80E-09	1.65E-16	3.96E-15	8.81E-15	5.85E-14	6.00E-09	2.5E-08	1.50E-08	9.34E-09	1.48E-09
In-119m*	1.20E-11	4.94E-13	1.97E-16	4.78E-15	1.13E-14	7.41E-14	0	0	0	1.82E-11	6.64E-13
Sn-113+	2.89E-09	5.08E-10	2.63E-16	1.16E-14	3.99E-16	2.30E-14	1.21E-09	5.12E-09	3.13E-10	3.96E-09	2.82E-10
Sn-126+	2.69E-08	1.31E-08	1.60E-15	7.20E-14	9.19E-15	1.31E-13	9.40E-09	3.80E-08	3.00E-09	1.55E-07	1.50E-08

	FGR-11 AND FGR-12 DCFs						RBE 30-DAY ACUTE DCFs			ICRP-60* DCFs	
	Inhalation (Sv/Bq)	Thyroid (Sv/Bq)	Ground (Sv/ (Bq s m^{-2}))	Immersion (Sv/ (Bq s m^{-3}))	Skin—GS (Sv/ (Bq s m^{-2}))	Skin—Air (Sv/ (Bq s m^2))	Colon (Gy-Eq/ Bq)	Lung (Gy-Eq/ Bq)	Red Marrow (Gy-Eq/ Bq)	Inhalation (Sv/Bq)	Thyroid (Sv/Bq)
Sb-124n*	2.80E-12	3.62E-13	4.30E-16	1.98E-14	1.53E-15	3.09E-14	0	0	0	5.98E-12	4.13E-13
Te-129m+	6.49E-09	3.96E-10	1.34E-16	3.51E-15	6.14E-15	3.90E-14	3.91E-09	1.06E-08	2.22E-09	7.95E-09	4.15E-09
Te-131m+	1.76E-09	3.67E-08	1.46E-15	7.12E-14	3.93E-15	1.04E-13	1.5E-09	1.51E-09	2.81E-10	1.08E-09	1.34E-08
Te-132+	2.65E-09	6.45E-08	2.39E-15	1.13E-13	7.76E-15	1.71E-13	3.43E-09	4.53E-09	5.71E-10	2.16E-09	2.64E-08
Te-133m+	1.20E-10	2.71E-09	1.91E-15	9.38E-14	6.08E-15	1.42E-13	0	0	0	9.14E-11	1.26E-09
Te-134+	6.99E-11	8.44E-10	3.32E-15	1.60E-13	1.05E-14	2.40E-13	1.40E-11	1.50E-11	1.90E-11	1.25E-10	4.96E-10
I-135+	3.32E-10	8.46E-09	1.53E-15	7.86E-14	4.76E-15	1.15E-13	3.00E-11	2.80E-11	2.70E-11	7.72E-10	1.30E-08
Xe-122*	0.00E+00	0.00E+00	1.10E-15	4.61E-14	1.25E-14	1.32E-13	0	0	0	0.00E+00	0.00E+00
Cs-137*	8.63E-09	7.93E-09	5.49E-16	2.55E-14	1.85E-15	4.40E-14	9.70E-10	7.60E-10	7.90E-10	3.92E-08	4.44E-09
Ba-126*	9.92E-11	7.59E-12	1.77E-15	7.85E-14	1.46E-14	1.88E-13	0	0	0	1.06E-10	6.18E-12
Ba-128*	8.20E-10	8.95E-11	1.01E-15	4.25E-14	1.06E-14	1.12E-13	0	0	0	1.36E-09	4.54E-11
Ce-134*	2.21E-09	4.79E-11	7.96E-16	3.30E-14	9.33E-15	9.29E-14	0	0	0	1.35E-09	8.31E-11
Ce-144*	1.01E-07	1.88E-09	1.75E-17	7.39E-16	2.44E-17	2.94E-15	9.70E-09	5.00E-08	8.90E-10	5.27E-08	5.56E-09
Ce-144+	1.01E-07	1.88E-09	1.78E-16	3.25E-15	1.27E-14	8.71E-14	9.70E-09	5.00E-08	8.90E-10	5.27E-08	5.56E-09
Nd-138*	2.78E-10	4.13E-12	9.38E-16	3.84E-14	1.25E-14	1.28E-13	0	0	0	2.48E-10	1.55E-11
Sm-142*	5.82E-11	1.15E-12	1.05E-15	4.3/E-14	1.3/E-14	1.47E-13	0	0	0	7.47E-11	5.88E-12
W-178*	7.32E-11	2.75E-12	1.28E-17	4.33E-16	1.83E-17	6.83E-16	3.71E-10	5.12E-11	1.81E-11	7.72E-10	2.53E-11
Os-180*	4.71E-12	8.89E-13	1.24E-15	5.91E-14	2.72E-15	8.02E-14	0	0	0	1.46E-11	1.04E-12
Pb-211*	2.35E-09	1.63E-10	2.08E-16	5.75E-15	9.90E-15	6.49E-14	0	0	0	1.20E-08	1.65E-10
Pb-212+	5.14E-08	3.58E-09	1.35E-15	7.16E-14	9.42E-15	1.36E-13	0	9.20E-09	1.12E-09	2.23E-07	3.58E-09
Bi-210m*	2.05E-06	1.00E-08	3.05E-16	1.18E-14	6.26E-15	5.04E-14	4.50E-09	1.20E-06	5.70E-10	9.90E-06	5.25E-09
Bi-212*	5.83E-09	1.64E-10	1.22E-15	6.55E-14	9.24E-15	1.22E-13	0	4.60E-09	1.80E-11	3.32E-08	1.80E-10
Bi-213*	4.63E-09	1.31E-10	2.08E-16	8.14E-15	4.47E-15	3.74E-14	0	0	0	3.20E-08	1.50E-10
Bi-214*	1.70E-09	5.07E-11	1.42E-15	7.11E-14	8.47E-15	1.27E-13	0	0	0	1.54E-08	6.59E-11

A-9

	FGR-11 AND FGR-12 DCFs						RBE 30-DAY ACUTE DCFs			ICRP-60* DCFs	
	Inhalation (Sv/Bq)	Thyroid (Sv/Bq)	Ground (Sv/(Bq s m^{-2}))	Immersion (Sv/(Bq s m^{-3}))	Skin—GS (Sv/(Bq s m^{-2}))	Skin—Air (Sv/(Bq s m^{-2}))	Colon (Gy-Eq/Bq)	Lung (Gy-Eq/Bq)	Red Marrow (Gy-Eq/Bq)	Inhalation (Sv/Bq)	Thyroid (Sv/Bq)
At-211*	2.76E-08	5.08E-09	3.51E-17	1.49E-15	5.87E-17	2.08E-15	0	0	0	1.19E-07	3.82E-09
Rn-222*	0.00E+00	0.00E+00	3.72E-19	1.73E-17	5.07E-19	2.23E-17	0	0	0	0.00E+00	0.00E+00
Rn-222+	3.89E-09	2.13E-10	1.66E-15	8.22E-14	9.38E-15	1.55E-13	0	0	0	3.01E-08	1.99E-10
Fr-222*	3.32E-09	3.29E-10	2.45E-16	8.61E-15	7.30E-15	5.12E-14	1.10E-12	2.80E-08	3.80E-10	2.78E-08	3.11E-10
Ra-223*	2.12E-06	3.38E-08	1.82E-16	8.31E-15	2.45E-16	1.28E-14	7.10E-09	3.00E-06	7.40E-09	8.67E-06	1.66E-08
Ra-223+	2.35E-09	1.63E-10	3.91E-16	1.41E-14	1.01E-14	7.77E-14	7.10E-09	3.00E-06	7.40E-09	8.68E-06	1.68E-08
Ra-224*	8.53E-07	1.53E-08	1.02E-17	4.80E-16	1.31E-17	7.02E-16	0	0	0	3.36E-06	1.23E-08
Ra-224+	9.04E-07	1.89E-08	1.36E-15	7.21E-14	9.44E-15	1.36E-13	3.90E-09	1.20E-06	4.30E-09	3.58E-06	1.59E-08
Ra-228+	1.37E-06	1.83E-07	8.40E-16	4.01E-14	4.41E-15	7.19E-14	3.90E-09	1.21E-06	5.42E-09	1.60E-05	2.07E-07
Ac-224*	3.56E-08	2.78E-12	2.06E-16	9.37E-15	2.94E-16	1.26E-14	1.14E-08	8.40E-06	4.61E-08	1.20E-07	5.90E-10
Ac-225*	2.92E-06	4.03E-11	4.03E-17	1.83E-15	6.67E-17	2.77E-15	4.30E-09	3.40E-06	1.20E-07	8.48E-06	4.37E-08
Ac-225+	2.92E-06	1.73E-10	2.51E-16	1.01E-14	4.83E-15	4.01E-14	4.30E-09	3.40E-06	1.20E-07	8.51E-06	4.39E-08
Ac-227*	1.81E-03	3.59E-08	1.30E-18	3.33E-17	4.18E-17	3.36E-16	1.00E-08	8.50E-06	6.50E-08	1.56E-04	2.29E-05
Ac-227+	1.81E-03	8.88E-08	5.00E-16	1.90E-14	1.02E-14	8.47E-14	1.88E-08	1.51E-05	8.96E-08	1.75E-04	2.30E-05
Th-226*	9.45E-09	1.62E-10	1.67E-17	7.63E-16	2.65E-17	1.23E-15	0	0	0	6.15E-08	6.46E-10
Th-227+	4.37E-06	5.37E-08	5.06E-16	1.93E-14	1.03E-14	8.56E-14	8.95E-09	6.72E-06	2.49E-08	1.91E-05	9.08E-08
Th-228+	9.32E-05	1.36E-06	1.36E-15	7.22E-14	9.45E-15	1.37E-13	6.71E-09	8.32E-06	4.49E-08	4.33E-05	3.27E-06
Th-234*	9.47E-09	1.03E-10	1.20E-16	1.74E-15	9.34E-15	5.54E-14	6.10E-09	2.40E-08	8.40E-10	7.69E-09	3.10E-10
Th-234+	9.47E-09	1.03E-10	1.25E-16	1.96E-15	9.35E-15	5.57E-14	6.10E-09	2.40E-08	8.40E-10	7.69E-09	3.10E-10
Pa-227*	1.32E-08	5.63E-14	7.93E-17	3.55E-15	1.39E-16	5.29E-15	0	0	0	7.94E-08	9.24E-10
U-240*	6.13E-10	2.37E-11	3.81E-16	1.51E-14	7.61E-15	6.26E-14	0	0	0	5.82E-10	1.03E-11
Pu-244+	1.09E-04	1.82E-08	4.00E-16	1.61E-14	7.75E-15	6.46E-14	4.70E-09	1.32E-06	1.24E-08	1.12E-04	6.64E-06
Cm-247+	1.12E-04	1.45E-08	3.21E-16	1.48E-14	4.86E-16	2.60E-14	1.50E-09	1.20E-06	6.90E-09	9.00E-05	7.04E-06
Es-254+	1.11E-05	2.44E-09	8.61E-16	4.19E-14	2.53E-15	6.57E-14	1.20E-07	2.50E-05	1.10E-06	1.02E-05	6.07E-09

	FGR-11 AND FGR-12 DCFs						RBE 30-DAY ACUTE DCFs			ICRP-60* DCFs	
	Inhalation (Sv/Bq)	Thyroid (Sv/Bq)	Ground (Sv/ (Bq s m^{-2}))	Immersion (Sv/ (Bq s m^{-3}))	Skin—GS (Sv/ (Bq s m^{-2}))	Skin—Air (Sv/ (Bq s m^{-2}))	Colon (Gy-Eq/ Bq)	Lung (Gy-Eq/ Bq)	Red Marrow (Gy-Eq/ Bq)	Inhalation (Sv/Bq)	Thyroid (Sv/Bq)
U-Natrl	3.20E-05	2.22E-08	1.43E-18	5.13E-17	8.03E-18	9.18E-17	1.00E-10	1.30E-06	1.00E-08	8.05E-06	3.00E-07
U-Enrch	3.21E-05	2.23E-08	7.82E-18	3.47E-16	1.66E-17	4.84E-16	6.70E-09	1.20E-06	1.00E-08	8.07E-06	3.01E-07
UF$_6$	6.63E-07	2.23E-08	7.82E-18	3.47E-16	1.66E-17	4.84E-16	1.00E-10	1.30E-06	1.00E-08	5.05E-07	3.01E-07

Table A-3 Short-Lived Radionuclides Not Included Explicitly in RASCAL 4

ISOTOPE	HALF-LIFE	COMMENT
N-13	9.965 min	
O-15	122.24 s	
Ne-19	17.22 s	
Al-28	1.140 min	Included as implicit daughter of Mg-28*
P-30	2.499 min	
K-38	7.636 min	
Ca-49	8.716 min	
Cu-62	9.74 min	Included as implicit daughter of Zn-62*
Cu-66	5.10 min	Included as implicit daughter of Ni-66*
Se-77m	17.45 s	
Kr-81m	13 s	
Rb-80	34 s	Included as implicit daughter of Sr-80*
Rb-82	1.3 min	Included as implicit daughter of Sr-82*
Nb-97m	60 s	Included as implicit daughter of Zr-97*
Rh-106	29.9 s	Included as implicit daughter of Ru-106*
Ag-108	2.37 min	Included as implicit daughter of Ag-108m*
Ag-109m	39.6 s	
Ag-110	24.6 s	Included as implicit daughter of Ag-110m*
In-114	71.9 s	Included as implicit daughter of In-114m*
In-119	2.4 min	Included as implicit daughter of In-119m*
Sb-124m	93 s	Included as implicit daughter of Sb-124n*
Sb-128m	10.4 min	
I-122	3.62 min	Included as implicit daughter of Xe-122*
Cs-126	1.64 min	Included as implicit daughter of Ba-126*
Cs-128	3.9 m	Included as implicit daughter of Ba-128*
Ba-137m	2.552 min	Included as implicit daughter of Cs-137*
La-134	6.67 min	Included as implicit daughter of Ce-134*
Pr-138	1.45 min	Included as implicit daughter of Nd-138*
Pr-144m	7.2 min	Included as implicit daughter of Ce-144*
Nd-141m	62.4 s	
Pm-142	40.5 s	Included as implicit daughter of Sm-142*
Ta-178m	9.31 min	Included as implicit daughter of W-178*
Re-180	2.43 min	Included as implicit daughter of Os-180*
Os-190m	9.9 min	
Ir-191m	4.94 s	
Au-195m	30.5 s	
Tl-206	4.20 min	Included as implicit daughter of Bi-210m*
Tl-207	4.77 min	Included as implicit daughter of Pb-211* and Pa-227*
Tl-208	3.07 min	Included as implicit daughter of Bi-212*
Tl-209	2.20 min	Included as implicit daughter of Bi-213*
Bi-211	2.14 min	Included as implicit daughter of Pb-211* and Pa-227*
Po-211	0.52 s	Included as implicit daughter of Pb-211*, At-211*, Pa-227*
Po-212	0.305 μs	Included as implicit daughter of Bi-212*
Po-213	4. 2 μs	Included as implicit daughter of Bi-213*
Po-214	164 μs	Included as implicit daughter of Bi-214*, Fr-222*, Th-226*
Po-215	1.78 ms	Included as implicit daughter of Ra-223*
Po-216	0.15 s	Included as implicit daughter of Ra-224*
Po-218	3.05 min	Included as implicit daughter of Rn-222*
At-215	0.1 ms	Included as implicit daughter of Pa-227*
At-216	0.3 ms	Included as implicit daughter of Ac-224*

ISOTOPE	HALF-LIFE	COMMENT
At-217	0.0323 s	Included as implicit daughter of Ac-225*
At-218	2 s	
Rn-218	35 ms	Included as implicit daughter of Th-226* and Fr-222*
Rn-219	3.96 s	Included as implicit daughter of Ra-223*
Rn-220	55.6 s	Included as implicit daughter of Ra-224*
Fr-219	21 ms	Included as implicit daughter of Pa-227*
Fr-220	27.4 s	Included as implicit daughter of Ac-224*
Fr-221	4.8 min	Included as implicit daughter of Ac-225*
Ra-222	38.0 s	Included as implicit daughter of Th-226* and Fr-222*
Ac-223	2.2 min	Incluced as implicit daughter of Pa-227*
Pa-234m	1.17 min	Incluced as implicit daughter of Th-234*
Np-240m	7.4 min	Incluced as implicit daughter of U-240*

Table A-4 RASCAL 4 Radionuclide Decay Chains

Key to the abbreviated table column headers:

Depos Type – Deposition type
Parent HL (days) – half-life of the parent in days
No. of Dau – the number of daughters in the simplified chain
D# - indicates a daughter where the # is replaced by a 1, 2, or 3; e.g. D1 name is the name of daughter 1
PI – parent index
HL – half-life
BF – branching fraction

(see the last paragraph of section A.3 for full discussion)

Parent Name	Depos Type	Parent HL (days)	No. of Dau	D1 Name	D1 PI	D1 HL (days)	D1 BF	D2 Name	D2 PI	D2 HL (days)	D2 BF	D2 Name	D3 PI	D3 HL (days)	D3 BF
H-3	0	4.51E-03	0												
Be-7	2	5.33E+01	0												
Be-10	2	5.84E+08	0												
C-11	2	1.42E-02	0												
C-14	2	2.09E+06	0												
F-18	1	7.62E-02	0												
Na-22	2	9.50E+02	0												
Na-24	2	6.25E-01	0												
Mg-28*	2	8.71E-01	0												
Al-26	2	2.61E+08	0												
Si-31	2	1.09E-01	0												
Si-32	2	4.82E+04	1	P-32	1	1.43E+01	1								
P-32	2	1.43E+01	0												
P-33	2	2.54E+01	0												
S-35	2	8.74E+01	0												
Cl-36	1	1.10E+08	0												
Cl-38	1	2.58E-02	0												
Cl-39	1	3.86E-02	1	Ar-39	1	9.82E+04	1								
Ar-37	0	3.50E+01	0												
Ar-39	0	9.82E+04	0												
Ar-41	0	7.61E-02	0												
K-40	2	4.67E+11	0												
K-42	2	5.15E-01	0												

A-14

Parent Name	Depos Type	Parent HL (days)	No. of Dau	D1 Name	D1 PI	D1 HL (days)	D1 BF	D2 Name	D2 PI	D2 HL (days)	D2 BF	D2 Name	D3 PI	D3 HL (days)	D3 BF
K-43	2	9.42E-01	0												
K-44	2	1.54E-02	0												
K-45	2	1.39E-02	1	Ca-45	1	1.63E+02	1								
Ca-41	2	3.72E+07	0												
Ca-45	2	1.63E+02	0												
Ca-47	2	4.53E+00	1	Sc-47	1	3.35E+00	1								
Sc-43	2	1.62E-01	0												
Sc-44m	2	2.44E+00	1	Sc-44	1	1.64E-01	0.9863								
Sc-44	2	1.64E-01	0												
Sc-46	2	8.38E+01	0												
Sc-47	2	3.35E+00	0												
Sc-48	2	1.82E+00	0												
Sc-49	2	3.99E-02	0												
Ti-44	2	2.19E+04	1	Sc-44	1	1.64E-01	1								
Ti-45	2	1.28E-01	0												
V-47	2	2.26E-02	0												
V-48	2	1.62E+01	0												
V-49	2	3.30E+02	0												
Cr-48	2	9.57E-01	1	V-48	1	1.62E+01	1								
Cr-49	2	2.92E-02	1	V-49	1	3.30E+02	1								
Cr-51	2	2.77E+01	0												
Mn-51	2	3.21E-02	1	Cr-51	1	2.77E+01	1								
Mn-52m	2	1.47E-02	1	Mn-52	1	5.59E+00	0.0175								
Mn-52	2	5.59E+00	0												
Mn-53	2	1.35E+09	0												
Mn-54	2	3.13E+02	0												
Mn-56	2	1.07E-01	0												
Fe-52	2	3.45E-01	1	Mn-52m	1	1.47E-02	1								
Fe-55	2	9.86E+02	0												
Fe-59	2	4.45E+01	0												
Fe-60	2	5.48E+08	1	Co-60m	1	7.27E-03	1								
Co-55	2	7.31E-01	1	Fe-55	1	9.86E+02	1								
Co-56	2	7.88E+01	0												
Co-57	2	2.71E+02	0												
Co-58m	2	3.81E-01	1	Co-58	1	7.08E+01	1								
Co-58	2	7.08E+01	0												

Parent Name	Depos Type	Parent HL (days)	No. of Dau	D1 Name	D1 PI	D1 HL (days)	D1 BF	D2 Name	D2 PI	D2 HL (days)	D2 BF	D2 Name	D3 PI	D3 HL (days)	D3 BF
Co-60m	2	7.27E-03	1	Co-60	1	1.92E+03	0.9975								
Co-60	2	1.92E+03	0												
Co-61	2	6.88E-02	0												
Co-62m	2	9.66E-03	0												
Ni-56	2	6.10E+00	1	Co-56	1	7.88E+01	1								
Ni-57	2	1.50E+00	1	Co-57	1	2.71E+02	1								
Ni-59	2	2.74E+07	0												
Ni-63	2	3.50E+04	0												
Ni-65	2	1.05E-01	0												
Ni-66*	2	2.28E+00	0												
Cu-60	2	1.61E-02	0												
Cu-61	2	1.42E-01	0												
Cu-64	2	5.29E-01	0												
Cu-67	2	2.58E+00	0												
Zn-62*	2	3.86E-01	0												
Zn-63	2	2.65E-02	0												
Zn-65	2	2.44E+02	0												
Zn-69m	2	5.73E-01	1	Zn-69	1	3.96E-02	0.9997								
Zn-69m+	2	5.73E-01	0												
Zn-69	2	3.96E-02	0												
Zn-71m	2	1.63E-01	0												
Zn-72	2	1.94E+00	1	Ga-72	1	5.88E-01	1								
Ga-65	2	1.06E-02	1	Zn-65	1	2.44E+02	1								
Ga-66	2	3.92E-01	0												
Ga-67	2	3.26E+00	0												
Ga-68	2	4.72E-02	0												
Ga-70	2	1.47E-02	0												
Ga-72	2	5.88E-01	0												
Ga-73	2	2.05E-01	0												
Ge-66	2	9.46E-02	1	Ga-66	1	3.92E-01	1								
Ge-67	2	1.30E-02	1	Ga-67	1	3.26E+00	1								
Ge-68	2	2.88E+02	1	Ga-68	1	4.72E-02	1								
Ge-68+	2	2.88E+02	0												
Ge-69	2	1.63E+00	0												
Ge-71	2	1.18E+01	0												
Ge-75	2	5.75E-02	0												

Parent Name	Depos Type	Parent HL (days)	No. of Dau	D1 Name	D1 PI	D1 HL (days)	D1 BF	D2 Name	D2 PI	D2 HL (days)	D2 BF	D2 Name	D3 PI	D3 HL (days)	D3 BF
Ge-77	2	4.71E-01	1	As-77	1	1.62E+00	1								
Ge-78	2	6.04E-02	1	As-78	1	6.30E-02	1								
As-69	2	1.06E-02	1	Ge-69	1	1.63E+00	1								
As-70	2	3.65E-02	0												
As-71	2	2.70E+00	1	Ge-71	1	1.18E+01	1								
As-72	2	1.08E+00	0												
As-73	2	8.03E+01	0												
As-74	2	1.78E+01	0												
As-76	2	1.10E+00	0												
As-77	2	1.62E+00	0												
As-78	2	6.30E-02	0												
Se-70	2	2.85E-02	1	As-70	1	3.65E-02	1								
Se-73m	2	2.71E-02	2	As-73	1	8.03E+01	0.27	Se-73	1	2.98E-01	1				
Se-73	2	2.98E-01	1	As-73	1	8.03E+01	1								
Se-75	2	1.20E+02	0												
Se-79	2	1.00E+00	0												
Se-81m	2	3.98E-02	1	Se-81	1	1.28E-02	1								
Se-81	2	1.28E-02	0												
Se-83	2	1.56E-02	1	Br-83	1	9.96E-02	1								
Br-74m	1	2.28E-02	0												
Br-74	1	1.76E-02	0												
Br-75	1	6.81E-02	1	Se-75	1	1.20E+02	1								
Br-76	1	6.75E-01	0												
Br-77	1	2.33E+00	0												
Br-80m	1	1.04E-01	1	Br-80	1	1.21E-02	1								
Br-80	1	1.21E-02	0												
Br-82	1	1.47E+00	0												
Br-83	1	9.96E-02	1	Kr-83m	1	7.63E-02	1								
Br-83+	1	9.96E-02	0												
Br-84	1	2.21E-02	0												
Kr-74	0	7.99E-03	1	Br-74	1	1.76E-02	1								
Kr-76	0	6.17E-01	1	Br-76	1	6.75E-01	1								
Kr-77	0	5.19E-02	1	Br-77	1	2.33E+00	1								
Kr-79	0	1.46E+00	0												
Kr-81	0	7.67E+07	0												
Kr-83m	0	7.63E-02	0												

Parent Name	Depos Type	Parent HL (days)	No. of Dau	D1 Name	D1 PI	D1 HL (days)	D1 BF	D2 Name	D2 PI	D2 HL (days)	D2 BF	D2 Name	D3 PI	D3 HL (days)	D3 BF
Kr-85m	0	1.87E-01	0												
Kr-85	0	3.91E+03	0												
Kr-87	0	5.30E-02	0												
Kr-88	0	1.18E-01	1	Rb-88	1	1.24E-02	1								
Rb-79	2	1.59E-02	1	Kr-79	1	1.46E+00	1								
Rb-81m	2	2.22E-02	1	Rb-81	1	1.91E-01	1								
Rb-81	2	1.91E-01	0												
Rb-82m	2	2.58E-01	0												
Rb-83	2	8.62E+01	1	Kr-83m	1	7.63E-02	0.762								
Rb-84	2	3.28E+01	0												
Rb-86	2	1.87E+01	0												
Rb-87	2	1.72E+13	0												
Rb-88	2	1.24E-02	0												
Rb-89	2	1.06E-02	1	Sr-89	1	5.05E+01	1								
Sr-80*	2	6.94E-02	0												
Sr-81	2	1.77E-02	1	Rb-81	1	1.91E-01	1								
Sr-82*	2	2.50E+01	0												
Sr-83	2	1.35E+00	1	Rb-83	1	8.62E+01	1								
Sr-85m	2	4.83E-02	1	Sr-85	1	6.48E+01	1								
Sr-85	2	6.48E+01	0												
Sr-87m	2	1.17E-01	0												
Sr-89	2	5.05E+01	0												
Sr-90	2	1.06E+04	1	Y-90	1	2.67E+00	1								
Sr-90+	2	1.06E+04	0												
Sr-91	2	3.96E-01	2	Y-91m	1	3.45E-02	0.578	Y-91	1	5.85E+01	1				
Sr-91+	2	3.96E-01	1	Y-91	1	5.85E+01	1								
Sr-92	2	1.13E-01	1	Y-92	1	1.48E-01	1								
Y-86m	2	3.33E-02	1	Y-86	1	6.14E-01	0.9931								
Y-86	2	6.14E-01	0												
Y-87	2	3.35E+00	1	Sr-87m	1	1.17E-01	0.999								
Y-88	2	1.07E+02	0												
Y-90m	2	1.33E-01	1	Y-90	1	2.67E+00	0.992								
Y-90	2	2.67E+00	0												
Y-91m	2	3.45E-02	1	Y-91	1	5.85E+01	1								
Y-91	2	5.85E+01	0												
Y-92	2	1.48E-01	0												

Parent Name	Depos Type	Parent HL (days)	No. of Dau	D1 Name	D1 PI	D1 HL (days)	D1 BF	D2 Name	D2 PI	D2 HL (days)	D2 BF	D2 Name	D3 PI	D3 HL (days)	D3 BF
Y-93	2	4.21E-01	0												
Y-94	2	1.33E-02	0												
Y-95	2	7.43E-03	1	Zr-95	1	6.40E+01	1								
Zr-86	2	6.88E-01	1	Y-86	1	6.14E-01	1								
Zr-88	2	8.34E+01	1	Y-88	1	1.07E+02	1								
Zr-89	2	3.27E+00	0												
Zr-93	2	5.58E+08	0												
Zr-95	2	6.40E+01	1	Nb-95	1	3.52E+01	1								
Zr-97*	2	7.04E-01	1	Nb-97	1	5.01E-02	0.053								
Zr-97+	2	7.04E-01	0												
Nb-88	2	9.93E-03	1	Zr-88	1	8.34E+01	1								
Nb-89m	2	4.58E-02	1	Zr-89	1	3.27E+00	1								
Nb-89	2	8.47E-02	1	Zr-89	1	3.27E+00	1								
Nb-90	2	6.08E-01	0												
Nb-93m	2	4.96E+03	0												
Nb-94	2	7.41E+06	0												
Nb-95m	2	3.61E+00	1	Nb-95	1	3.52E+01	1								
Nb-95	2	3.52E+01	0												
Nb-96	2	9.73E-01	0												
Nb-97	2	5.01E-02	0												
Nb-98	2	3.58E-02	0												
Mo-90	2	2.36E-01	1	Nb-90	1	6.08E-01	1								
Mo-93m	2	2.85E-01	1	Mo-93	1	1.28E+06	1								
Mo-93	2	1.28E-06	1	Nb-93m	1	4.96E+03	1								
Mo-99	2	2.75E+00	1	Tc-99m	1	2.51E-01	0.876								
Mo-99+	2	2.75E+00	0												
Mo-101	2	1.02E-02	1	Tc-101	1	9.86E-03	1								
Tc-93m	2	3.02E-02	1	Mo-93	1	1.28E+06	0.182	Tc-93	1	1.15E-01	0.818				
Tc-93	2	1.15E-01	1	Mo-93	1	1.28E+06	1								
Tc-94m	2	3.61E-02	0												
Tc-94	2	2.03E-01	0												
Tc-95m	2	6.10E+01	1	Tc-95	1	8.33E-01	0.04								
Tc-95	2	8.33E-01	0												
Tc-96m	2	3.58E-02	1	Tc-96	1	4.28E+00	0.98								
Tc-96	2	4.28E+00	0												
Tc-97m	2	8.70E+01	1	Tc-97	1	9.49E+08	1								

Parent Name	Depos Type	Parent HL (days)	No. of Dau	D1 Name	D1 PI	D1 HL (days)	D1 BF	D2 Name	D2 PI	D2 HL (days)	D2 BF	D3 Name	D3 PI	D3 HL (days)	D3 BF
Tc-97	2	9.49E+08	0												
Tc-98	2	1.53E+09	0												
Tc-99m	2	2.51E-01	0												
Tc-99	2	7.77E+07	0												
Tc-101	2	9.86E-03	0												
Tc-104	2	1.26E-02	0												
Ru-94	2	3.60E-02	1	Tc-94m	1	3.61E-02	1								
Ru-97	2	2.90E+00	2	Tc-97m	1	8.70E+01	0.0008	Tc-97	1	9.49E+08	0.9992				
Ru-103	2	3.93E+01	1	Rh-103m	1	3.90E-02	0.997								
Ru-103+	2	3.93E+01	0												
Ru-105	2	1.85E-01	1	Rh-105	1	1.47E+00	1								
Ru-106*	2	3.68E-02	0												
Rh-99m	2	1.96E-01	0												
Rh-99	2	1.60E+01	0												
Rh-100	2	8.67E-01	0												
Rh-101m	2	4.34E+00	1	Rh-101	1	1.17E+03	0.072								
Rh-101	2	1.17E+03	0												
Rh-102m	2	1.37E+03	1	Rh-102	1	1.06E+03	0.05								
Rh-102	2	1.06E-03	0												
Rh-103m	2	3.90E-02	0												
Rh-105	2	1.47E+00	0												
Rh-106m	2	9.17E-02	0												
Rh-107	2	1.51E-02	0												
Pd-100	2	3.63E+00	1	Rh-100	1	8.67E-01	1								
Pd-101	2	3.45E-01	2	Rh-101m	1	4.34E+00	0.997	Rh-101	1	1.17E+03	0.003				
Pd-103	2	1.70E+01	1	Rh-103m	1	3.90E-02	1								
Pd-103+	2	1.70E+01	0												
Pd-107	2	2.37E-09	0												
Pd-109	2	5.59E-01	0												
Ag-102	2	8.96E-03	0												
Ag-103	2	4.56E-02	2	Pd-103	1	1.70E+01	1	Rh-103m	2	3.90E-02	1				
Ag-104m	2	2.33E-02	1	Ag-104	1	4.81E-02	0.33								
Ag-104	2	4.81E-02	0												
Ag-105	2	4.10E+01	0												
Ag-106m	2	8.41E+00	0												
Ag-106	2	1.66E-02	0												

Parent Name	Depos Type	Parent HL (days)	No. of Dau	D1 Name	D1 PI	D1 HL (days)	D1 BF	D2 Name	D2 PI	D2 HL (days)	D2 BF	D2 Name	D3 PI	D3 HL (days)	D3 BF
Ag-108m*	2	1.53E+05	0												
Ag-110m*	2	2.50E+02	0												
Ag-111	2	7.45E+00	0												
Ag-112	2	1.30E-01	0												
Ag-115	2	1.39E-02	3	Cd-115	1	2.23E+00	0.934	In-115m	2	1.87E-01	1	Cd-115m	1	4.46E+01	0.066
Cd-104	2	4.01E-02	1	Ag-104	1	4.81E-02	1								
Cd-107	2	2.70E-01	0												
Cd-109	2	4.64E+02	0												
Cd-113m	2	4.96E+03	0												
Cd-113	2	3.39E+18	0												
Cd-115m	2	4.46E+01	0												
Cd 116	2	2.23E+00	1	In 115m	1	1.87E-01	1								
Cd-115+	2	2.23E+00	0												
Cd-117m	2	1.40E-01	3	In-117	1	3.04E-02	1	Sn-117m	2	1.36E+01	0.0032	In-117m	1	8.09E-02	0.01
Cd-117	2	1.04E-01	3	In 117	1	3.04E-02	0.5124	Sn-117m	2	1.36E-01	0.0032	In-117m	1	8.09E-02	0.92
In-109	2	1.75E-01	1	Cd-109	1	4.64E+02	1								
In-110	2	2.04E-01	0												
In-110m	2	4.80E-02	0												
In-111	2	2.83E+00	0												
In-112	2	1.00E-02	0												
In-113m	2	6.91E-02	0												
In-114m*	2	4.95E+01	0												
In-115m	2	1.87E-01	0												
In-115	2	1.61E+17	0												
In-116m	2	3.76E-02	0												
In-117m	2	8.09E-02	2	In-117	1	3.04E-02	0.471	Sn-117m	2	1.36E+01	0.0032				
In-117	2	3.04E-02	1	Sn-117m	1	1.36E+01	0.0032								
In-119m*	2	1.25E-02	1	Sn-119m	1	2.93E+02	0.0027								
Sn-110	2	1.67E-01	1	In-110a	1	4.80E-02	1								
Sn-111	2	2.45E-02	1	In-111	1	2.83E+00	1								
Sn-113	2	1.15E+02	1	In-113m	1	6.91E-02	1								
Sn-113+	2	1.15E+02	0												
Sn-117m	2	1.36E+01	0												
Sn-119m	2	2.93E+02	0												
Sn-121m	2	2.01E+04	1	Sn-121	1	1.13E+00	0.770								
Sn-121	2	1.13E+00	0												

Parent Name	Depos Type	Parent HL (days)	No. of Dau	D1 Name	D1 PI	D1 HL (days)	D1 BF	D2 Name	D2 PI	D2 HL (days)	D2 BF	D2 Name	D3 PI	D3 HL (days)	D3 BF
Sn-123m	2	2.78E-02	0												
Sn-123	2	1.29E+02	0												
Sn-125	2	9.64E+00	2	Sb-125	1	1.01E+03	1	Te-125m	2	5.80E+01	0.228				
Sn-126	2	8.40E+07	2	Sb-126m	1	1.32E-02	0.14	Sb-126	2	1.24E+01	0.14				
Sn-126+	2	8.40E+07	1	Sb-126	1	1.24E+01	0.14								
Sn-127	2	8.75E-02	2	Sb-127	1	3.85E+00	1	Te-127	2	3.90E-01	0.824	Te-127m	2	1.09E-02	0.176
Sn-128	2	4.10E-02	1	Sb-128a	1	7.22E-03	1								
Sb-115	2	2.21E-02	0												
Sb-116m	2	4.19E-02	0												
Sb-116	2	1.10E-02	0												
Sb-117	2	1.17E-01	0												
Sb-118m	2	2.08E-01	0												
Sb-119	2	1.59E+00	0												
Sb-120m	2	5.76E+00	0												
Sb-120	2	1.10E-02	0												
Sb-122	2	2.70E+00	0												
Sb-124n*	2	1.40E-02	1	Sb-124	1	6.02E+01	0.8								
Sb-124	2	6.02E+01	0												
Sb-125	2	1.01E+03	1	Te-125m	1	5.80E+01	0.228								
Sb-126m	2	1.32E-02	1	Sb-126	1	1.24E+01	0.14								
Sb-126	2	1.24E+01	0												
Sb-127	2	3.85E+00	2	Te-127m	1	1.09E+02	0.176	Te-127	1	3.90E-01	1				
Sb-128m	2	7.22E-03	0												
Sb-128	2	3.75E-01	0												
Sb-129	2	1.80E-01	2	Te-129m	1	3.36E+01	0.225	Te-129	2	4.83E-02	0.775				
Sb-130	2	2.78E-02	0												
Sb-131	2	1.60E-02	3	Te-131	1	1.74E-02	1	I-131	2	8.04E+00	1	Te-131m	1	1.25E+00	0.0993
Te-116	2	1.04E-01	1	Sb-116	1	1.10E-02	1								
Te-121m	2	1.54E+02	1	Te-121	1	1.70E+01	0.886								
Te-121	2	1.70E+01	0												
Te-123m	2	1.20E+02	0												
Te-123	2	2.19E+17	0												
Te-125m	2	5.80E+01	0												
Te-127m	2	1.09E+02	1	Te-127	1	3.90E-01	0.976								
Te-127	2	3.90E-01	0												
Te-129m	2	3.36E+01	1	Te-129	1	4.83E-02	0.65								

Parent Name	Depos Type	Parent HL (days)	No. of Dau	D1 Name	D1 PI	D1 HL (days)	D1 BF	D2 Name	D2 PI	D2 HL (days)	D2 BF	D2 Name	D3 PI	D3 HL (days)	D3 BF
Te-129m+	2	3.36E-01	0												
Te-129	2	4.83E-02	0												
Te-131m	2	1.25E+00	2	Te-131	1	1.74E-02	0.222	I-131	1	8.04E+00	1				
Te-131m+	2	1.25E+00	1	I-131	1	8.04E+00	1								
Te-131	2	1.74E-02	1	I-131	1	8.04E+00	1								
Te-132	2	3.26E+00	1	I-132	1	9.58E-02	1								
Te-132+	2	3.26E+00	0												
Te-133m	2	3.85E-02	2	I-133	1	8.67E-01	0.87	Te-133	1	8.65E-03	0.13				
Te-133m+	2	3.85E-02	1	I-133	1	8.67E-01	1								
Te-133	2	8.65E-03	1	I-133	1	8.67E-01	1								
Te-134	2	2.90E-02	1	I-134	1	3.65E-02	1								
To-134+	2	2.00E 02	0												
I-120m	1	3.68E-02	0												
I-120	1	5.63E-02	0												
I-121	1	8.83E 02	1	Tc 121	1	1.70E 01	1								
I-123	1	5.50E-01	1	Te-123m	1	1.20E+02	0.0001								
I-124	1	4.18E+00	0												
I-125	1	6.01E+01	0												
I-126	1	1.30E+01	0												
I-128	1	1.74E-02	0												
I-129	1	5.73E+09	0												
I-130	1	5.15E-01	0												
I-131	1	8.04E+00	0												
I-132m	1	5.81E-02	1	I-132	1	9.58E-02	0.86								
I-132	1	9.58E-02	0												
I-133	1	8.67E-01	0												
I-134	1	3.65E-02	0												
I-135	1	2.75E-01	2	Xe-135m	1	1.06E-02	0.154	Xe-135	1	3.79E-01	0.846				
I-135+	1	2.75E-01	1	Xe-135	1	3.79E-01	0.846								
Xe-120	0	2.78E-02	1	I-120	1	5.63E-02	1								
Xe-121	0	2.78E-02	2	I-121	1	8.83E-02	1	Te-121	2	1.70E+01	1				
Xe-122*	0	8.38E-01	0												
Xe-123	0	8.67E-02	1	I-123	1	5.50E-01	1								
Xe-125	0	7.08E-01	1	I-125	1	6.01E+01	1								
Xe-127	0	3.64E+01	0												
Xe-129m	0	8.00E+00	0												

Parent Name	Depos Type	Parent HL (days)	No. of Dau	D1 Name	D1 PI	D1 HL (days)	D1 BF	D2 Name	D2 PI	D2 HL (days)	D2 BF	D2 Name	D3 PI	D3 HL (days)	D3 BF
Xe-131m	0	1.19E+01	0												
Xe-133m	0	2.19E+00	1	Xe-133	1	5.25E+00	1								
Xe-133	0	5.25E+00	0												
Xe-135m	0	1.06E-02	1	Xe-135	1	3.79E-01	0.9999								
Xe-135	0	3.79E-01	0												
Xe-138	0	9.84E-03	1	Cs-138	1	2.24E-02	1								
Cs-125	2	3.13E-02	0												
Cs-127	2	2.60E-01	0												
Cs-129	2	1.34E+00	0												
Cs-130	2	2.08E-02	0												
Cs-131	2	9.69E+00	0												
Cs-132	2	6.48E+00	0												
Cs-134m	2	1.21E-01	1	Cs-134	1	7.53E+02	1								
Cs-134	2	7.53E+02	0												
Cs-135m	2	3.68E-02	0												
Cs-135	2	8.40E+08	0												
Cs-136	2	1.31E+01	0												
Cs-137*	2	1.10E+04	0												
Cs-138	2	2.24E-02	0												
Ba-126*	2	6.70E-02	0												
Ba-128*	2	2.43E+00	0												
Ba-131m	2	1.01E-02	2	Ba-131	1	1.18E+01	1	Cs-131	2	9.69E+00	1				
Ba-131	2	1.18E+01	1	Cs-131	1	9.69E+00	1								
Ba-133m	2	1.62E+00	1	Ba-133	1	3.92E+03	1								
Ba-133	2	3.92E+03	0												
Ba-135m	2	1.20E+00	0												
Ba-139	2	5.74E-02	0												
Ba-140	2	1.27E+01	1	La-140	1	1.68E+00	1								
Ba-141	2	1.27E-02	2	La-141	1	1.64E-01	1	Ce-141	2	3.25E+01	1				
Ba-142	2	7.36E-03	1	La-142	1	6.42E-02	1								
La-131	2	4.10E-02	2	Ba-131	1	1.18E+01	1	Cs-131	2	9.69E+00	1				
La-132	2	2.00E-01	0												
La-135	2	8.13E-01	0												
La-137	2	2.19E-07	0												
La-138	2	4.93E+13	0												
La-140	2	1.68E+00	0												

Parent Name	Depos Type	Parent HL (days)	No. of Dau	D1 Name	D1 PI	D1 HL (days)	D1 BF	D2 Name	D2 PI	D2 HL (days)	D2 BF	D2 Name	D3 PI	D3 HL (days)	D3 BF
La-141	2	1.64E-01	1	Ce-141	1	3.25E+01	1								
La-142	2	6.42E-02	0												
La-143	2	9.88E-03	2	Ce-143	1	1.38E+00	1	Pr-143	2	1.36E+01	1				
Ce-134*	2	3.00E+00	0												
Ce-135	2	7.33E-01	1	La-135	1	8.13E-01	1								
Ce-137m	2	1.43E+00	1	Ce-137	1	3.75E-01	0.9941								
Ce-137	2	3.75E-01	0												
Ce-139	2	1.38E+02	0												
Ce-141	2	3.25E+01	0												
Ce-143	2	1.38E+00	1	Pr-143	1	1.36E+01	1								
Ce-144*	2	2.84E+02	1	Pr-144	1	1.20E-02	1								
Ce-144+	2	2.84E+02	0												
Pr-136	2	9.10E-03	0												
Pr-137	2	5.32E-02	1	Ce-137	1	3.75E-01	1								
Pr-138m	2	8.75E-02	0												
Pr-139	2	1.88E-01	1	Ce-139	1	1.38E+02	1								
Pr-142m	2	1.01E-02	1	Pr-142	1	7.97E-01	1								
Pr-142	2	7.97E-01	0												
Pr-143	2	1.36E+01	0												
Pr-144	2	1.20E-02	0												
Pr-145	2	2.49E-01	0												
Pr-147	2	9.44E-03	1	Nd-147	1	1.10E+01	1								
Nd-136	2	3.53E-02	1	Pr-136	1	9.10E-03	1								
Nd-138*	2	2.10E-01	0												
Nd-139m	2	2.29E-01	3	Pr-139	1	1.88E-01	1	Ce-139	2	1.38E+02	1	Nd-139	1	2.06E-02	0.12
Nd-139	2	2.06E-02	2	Pr-139	1	1.88E-01	1	Ce-139	2	1.38E+02	1				
Nd-141	2	1.04E-01	0												
Nd-147	2	1.10E+01	1	Pm-147	1	9.58E+02	1								
Nd-149	2	7.21E-02	1	Pm-149	1	2.21E+00	1								
Nd-151	2	8.64E-03	2	Pm-151	1	1.18E+00	1	Sm-151	2	3.29E+04	1				
Pm-141	2	1.45E-02	1	Nd-141	1	1.04E-01	1								
Pm-143	2	2.65E+02	0												
Pm-144	2	3.63E+02	0												
Pm-145	2	6.46E+03	0												
Pm-146	2	2.02E+03	0												
Pm-147	2	9.58E+02	0												

Parent Name	Depos Type	Parent HL (days)	No. of Dau	D1 Name	D1 PI	D1 HL (days)	D1 BF	D2 Name	D2 PI	D2 HL (days)	D2 BF	D2 Name	D3 PI	D3 HL (days)	D3 BF
Pm-148m	2	4.13E+01	1	Pm-148	1	5.37E+00	0.046								
Pm-148	2	5.37E+00	0												
Pm-149	2	2.21E+00	0												
Pm-150	2	1.12E-01	0												
Pm-151	2	1.18E+00	0												
Sm-141m	2	1.57E-02	2	Pm-141	1	1.45E-02	1	Sm-141	2	7.08E-03	0.0003				
Sm-141	2	7.08E-03	1	Pm-141	1	1.45E-02	1								
Sm-142	2	5.03E-02	0												
Sm-142*	2	5.03E-02	0												
Sm-145	2	3.40E+02	1	Pm-145	1	6.46E+03	1								
Sm-146	2	3.76E+10	0												
Sm-147	2	3.87E+13	0												
Sm-151	2	3.29E+04	0												
Sm-153	2	1.95E+00	0												
Sm-155	2	1.53E-02	1	Eu-155	1	1.81E+03	1								
Sm-156	2	3.92E-01	1	Eu-156	1	1.52E+01	1								
Eu-145	2	5.94E+00	2	Sm-145	1	3.40E+02	1	Pm-145	2	6.46E+03	1				
Eu-146	2	4.61E+00	0												
Eu-147	2	2.40E+01	0												
Eu-148	2	5.45E+01	0												
Eu-149	2	9.31E+01	0												
Eu-150	2	1.25E+04	0												
Eu-150m	2	5.26E-01	0												
Eu-152m	2	3.88E-01	0												
Eu-152	2	4.87E+03	0												
Eu-154	2	3.21E+03	0												
Eu-155	2	1.81E+03	0												
Eu-156	2	1.52E+01	0												
Eu-157	2	6.31E-01	0												
Eu-158	2	3.19E-02	0												
Gd-145	2	1.59E-02	1	Eu-145	1	5.94E+00	1								
Gd-146	2	4.83E+01	1	Eu-146	1	4.61E+00	1								
Gd-147	2	1.59E+00	1	Eu-147	1	2.40E+01	1								
Gd-148	2	3.39E+04	0												
Gd-149	2	9.40E+00	1	Eu-149	1	9.31E+01	1								
Gd-151	2	1.20E+02	0												

Parent Name	Depos Type	Parent HL (days)	No. of Dau	D1 Name	D1 PI	D1 HL (days)	D1 BF	D2 Name	D2 PI	D2 HL (days)	D2 BF	D2 Name	D3 PI	D3 HL (days)	D3 BF
Gd-152	2	3.94E+16	0												
Gd-153	2	2.42E+02	0												
Gd-159	2	7.73E-01	0												
Tb-147	2	6.88E-02	2	Gd-147	1	1.59E+00	1	Eu-147	2	2.40E+01	1				
Tb-149	2	1.73E-01	3	Gd-149	1	9.40E+00	0.8	Eu-149	2	9.31E+01	1	Eu-145	1	5.94E+00	0.2
Tb-150	2	1.36E-01	0												
Tb-151	2	7.33E-01	1	Gd-151	1	1.20E+02	1								
Tb-153	2	2.34E+00	1	Gd-153	1	2.42E+02	1								
Tb-154	2	8.92E-01	0												
Tb-155	2	5.32E+00	0												
Tb-156m	2	1.02E+00	1	Tb-156	1	5.34E+00	1								
Tb-156n	2	2.08E-01	1	Tb-156	1	5.34E+00	1								
Tb-156	2	5.34E+00	0												
Tb-157	2	2.59E+04	0												
Tb-158	2	5.48E+04	0												
Tb-160	2	7.23E-01	0												
Tb-161	2	6.91E+00	0												
Dy-155	2	4.17E-01	1	Tb-155	1	5.32E+00	1								
Dy-157	2	3.38E-01	0												
Dy-159	2	1.44E+02	0												
Dy-165	2	9.73E-02	0												
Dy-166	2	3.40E+00	1	Ho-166	1	1.12E+00	1								
Ho-155	2	3.33E-02	2	Dy-155	1	4.17E-01	1	Tb-155	2	5.32E+00	1				
Ho-157	2	8.75E-03	1	Dy-157	1	3.38E-01	1								
Ho-159	2	2.29E-02	1	Dy-159	1	1.44E+02	1								
Ho-161	2	1.04E-01	0												
Ho-162m	2	4.72E-02	1	Ho-162	1	1.04E-02	0.61								
Ho-162	2	1.04E-02	0												
Ho-164m	2	2.60E-02	1	Ho-164	1	2.01E-02	1								
Ho-164	2	2.01E-02	0												
Ho-166m	2	4.38E+05	0												
Ho-166	2	1.12E+00	0												
Ho-167	2	1.29E-01	0												
Er-161	2	1.35E-01	1	Ho-161	1	1.04E-01	1								
Er 166	2	4.32E 01	0												
Er-169	2	9.30E+00	0												

Parent Name	Depos Type	Parent HL (days)	No. of Dau	D1 Name	D1 PI	D1 HL (days)	D1 BF	D2 Name	D2 PI	D2 HL (days)	D2 BF	D2 Name	D3 PI	D3 HL (days)	D3 BF
Er-171	2	3.13E-01	1	Tm-171	1	7.01E+02	1								
Er-172	2	2.05E+00	1	Tm-172	1	2.65E+00	1								
Tm-162	2	1.51E-02	0												
Tm-166	2	3.21E-01	0												
Tm-167	2	9.24E+00	0												
Tm-170	2	1.29E-02	0												
Tm-171	2	7.01E+02	0												
Tm-172	2	2.65E+00	0												
Tm-173	2	3.43E-01	0												
Tm-175	2	1.06E-02	1	Yb-175	1	4.19E+00	1								
Yb-162	2	1.31E-02	1	Tm-162	1	1.51E-02	1								
Yb-166	2	2.36E+00	1	Tm-166	1	3.21E-01	1								
Yb-167	2	1.22E-02	1	Tm-167	1	9.24E+00	1								
Yb-169	2	3.20E+01	0												
Yb-175	2	4.19E+00	0												
Yb-177	2	7.92E-02	1	Lu-177	1	6.71E+00	1								
Yb-178	2	5.14E-02	1	Lu-178	1	1.97E-02	1								
Lu-169	2	1.42E+00	1	Yb-169	1	3.20E+01	1								
Lu-170	2	2.00E+00	0												
Lu-171	2	8.22E+00	0												
Lu-172	2	6.70E+00	0												
Lu-173	2	5.00E+02	0												
Lu-174m	2	1.42E+02	1	Lu-174	1	1.21E+03	0.993								
Lu-174	2	1.21E+03	0												
Lu-176m	2	1.53E-01	0												
Lu-176	2	1.31E+13	0												
Lu-177m	2	1.61E+02	1	Lu-177	1	6.71E+00	0.21								
Lu-177	2	6.71E+00	0												
Lu-178m	2	1.58E-02	0												
Lu-178	2	1.97E-02	0												
Lu-179	2	1.91E-01	0												
Hf-170	2	6.67E-01	1	Lu-170	1	2.00E+00	1								
Hf-172	2	6.83E+02	1	Lu-172	1	6.70E+00	1								
Hf-173	2	1.00E+00	1	Lu-173	1	5.00E+02	1								
Hf-175	2	7.00E+01	0												
Hf-177m	2	3.57E-02	0												

Parent Name	Depos Type	Parent HL (days)	No. of Dau	D1 Name	D1 PI	D1 HL (days)	D1 BF	D2 Name	D2 PI	D2 HL (days)	D2 BF	D3 PI	D3 HL (days)	D3 BF
Hf-178m	2	1.13E+04	0											
Hf-179m	2	2.51E+01	0											
Hf-180m	2	2.29E-01	0											
Hf-181	2	4.24E+01	0											
Hf-182m	2	4.27E-02	1	Ta-182	1	1.15E+02	0.54							
Hf-182	2	3.29E+09	1	Ta-182	1	1.15E+02	1							
Hf-183	2	4.44E-02	1	Ta-183	1	5.10E+00	1							
Hf-184	2	1.72E-01	1	Ta-184	1	3.63E-01	1							
Ta-172	2	2.56E-02	2	Hf-172	1	6.83E+02	1	Lu-172	2	6.70E+00	1			
Ta-173	2	1.52E-01	2	Hf-173	1	1.00E+00	1	Lu-173	2	5.00E+02	1			
Ta-174	2	5.00E-02	0											
Ta-175	2	4.38E-01	1	Hf-175	1	7.00E+01	1							
Ta-176	2	3.37E-01	0											
Ta-177	2	2.36E+00	0											
Ta-178m	2	9.17E-02	0											
Ta-179	2	6.65E+02	0											
Ta-180m	2	3.38E-01	0											
Ta-180	2	3.65E+15	0											
Ta-182m	2	1.10E-02	1	Ta-182	1	1.15E+02	1							
Ta-182	2	1.15E+02	0											
Ta-183	2	5.10E+00	0											
Ta-184	2	3.63E-01	0											
Ta-185	2	3.40E-02	1	W-185	1	7.51E+01	1							
Ta-186	2	7.29E-03	0											
W-176	2	9.58E-02	1	Ta-176	1	3.37E-01	1							
W-177	2	9.38E-03	1	Ta-177	1	2.36E+00	1							
W-178	2	2.17E+01	0											
W-178*	2	2.17E+01	0											
W-179	2	2.60E-02	1	Ta-179	1	6.65E+02	1							
W-181	2	1.21E+02	0											
W-185	2	7.51E+01	0											
W-187	2	9.96E-01	0											
W-188	2	6.94E+01	1	Re-188	1	7.08E-01	1							
Re-177	2	9.72E-03	2	W-177	1	9.38E-03	1	Ta-177	2	2.36E+00	1			
Re-178	2	9.17E-03	1	W-178*	1	2.17E+01	1							
Re-181	2	8.33E-01	1	W-181	1	1.21E+02	1							

Parent Name	Depos Type	Parent HL (days)	No. of Dau	D1 Name	D1 PI	D1 HL (days)	D1 BF	D2 Name	D2 PI	D2 HL (days)	D2 BF	D3 Name	D3 PI	D3 HL (days)	D3 BF
Re-182	2	2.67E+00	0												
Re-182m	2	5.29E-01	0												
Re-184m	2	1.65E+02	1	Re-184	1	3.80E+01	0.747								
Re-184	2	3.80E+01	0												
Re-186m	2	7.30E+07	1	Re-186	1	3.78E+00	1								
Re-186	2	3.78E+00	0												
Re-187	2	1.83E+13	0												
Re-188m	2	1.29E-02	1	Re-188	1	7.07E-01	1								
Re-188	2	7.07E-01	0												
Re-189	2	1.01E+00	1	Os-189m	1	2.50E-01	0.241								
Os-180	2	1.53E-02	0												
Os-180*	2	1.53E-02	0												
Os-181	2	7.29E-02	2	Re-181	1	8.33E-01	1	W-181	2	1.21E+02	1				
Os-182	2	9.17E-01	2	Re-182a	1	5.29E-01	1								
Os-185	2	9.40E+01	0												
Os-189m	2	2.50E-01	0												
Os-191m	2	5.43E-01	1	Os-191	1	1.54E+01	1								
Os-191	2	1.54E+01	0												
Os-193	2	1.25E+00	0												
Os-194	2	2.19E+03	1	Ir-194	1	7.98E-01	1								
Ir-182	2	1.04E-02	2	Os-182	1	9.17E-01	1	Re-182a	2	5.29E-01	1				
Ir-184	2	1.26E-01	0												
Ir-185	2	5.83E-01	1	Os-185	1	9.40E+01	1								
Ir-186	2	6.58E-01	0												
Ir-186m	2	7.29E-02	0												
Ir-187	2	4.38E-01	0												
Ir-188	2	1.73E+00	0												
Ir-189	2	1.33E+01	1	Os-189m	1	2.50E-01	0.083								
Ir-190n	2	1.29E-01	2	Ir-190m	1	5.00E-02	0.05	Ir-190	2	1.21E+01	1				
Ir-190m	2	5.00E-02	1	Ir-190	1	1.21E+01	1								
Ir-190	2	1.21E+01	0												
Ir-192m	2	8.80E+04	1	Ir-192	1	7.40E+01	1								
Ir-192	2	7.40E+01	0												
Ir-194m	2	1.71E-02	0												
Ir-194	2	7.98E-01	0												
Ir-195m	2	1.58E-01	1	Ir-195	1	1.04E-01	0.04								

Parent Name	Depos Type	Parent HL (days)	No. of Dau	D1 Name	D1 PI	D1 HL (days)	D1 BF	D2 Name	D2 PI	D2 HL (days)	D2 BF	D2 Name	D3 PI	D3 HL (days)	D3 BF
Ir-195	2	1.04E-01	0												
Pt-186	2	8.33E-02	1	Ir-186b	1	7.29E-02	1								
Pt-188	2	1.02E-01	1	Ir-188	1	1.73E+00	1								
Pt-189	2	4.53E-01	2	Ir-189	1	1.33E+01	1	Os-189m	2	2.50E-01	0.083				
Pt-191	2	2.80E+00	0												
Pt-193m	2	4.33E+00	1	Pt-193	1	1.83E+04	1								
Pt-193	2	1.83E+04	0												
Pt-195m	2	4.02E+00	0												
Pt-197m	2	6.56E-02	1	Pt-197	1	7.63E-01	0.967								
Pt-197	2	7.63E-01	0												
Pt-199	2	2.14E-02	1	Au-199	1	3.14E+00	1								
Pt-200	2	5.21E-01	1	Au-200	1	3.36E-02	1								
Au-193	2	7.35E-01	1	Pt-193	1	1.83E+04	1								
Au-194	2	1.65E+00	0												
Au-195	2	1.83E+02	0												
Au-198m	2	2.30E+00	1	Au-198	1	2.70E+00	1								
Au-198	2	2.70E+00	0												
Au-199	2	3.14E+00	0												
Au-200m	2	7.79E-01	1	Au-200	1	3.36E-02	1								
Au-200	2	3.36E-02	0												
Au-201	2	1.83E-02	0												
Hg-193m	2	4.63E-01	3	Au-193	1	7.35E-01	1	Pt-193	2	1.83E+04	1	Hg-193	1	1.46E-01	0.08
Hg-193	2	1.46E-01	2	Au-193	1	7.35E-01	1	Pt-193	2	1.83E+04	1				
Hg-194	2	1.61E+05	1	Au-194	1	1.65E+00	1								
Hg-195m	2	1.73E+00	2	Au-195	1	1.03E-02	1	Hg-195	1	4.13E-01	0.542				
Hg-195	2	4.13E-01	1	Au-195	1	1.83E+02	1								
Hg-197m	2	9.92E-01	1	Hg-197	1	2.67E+00	0.93								
Hg-197	2	2.67E+00	0												
Hg-199m	2	2.96E-02	0												
Hg-203	2	4.66E+01	0												
Tl-194m	2	2.28E-02	1	Hg-194	1	1.61E+05	1								
Tl-194	2	2.29E-02	1	Hg-194	1	1.61E+05	1								
Tl-195	2	4.83E-02	2	Hg-195	1	4.13E-01	1	Au-195	2	1.83E+02	1				
Tl-197	2	1.18E-01	1	Hg-197	1	2.67E+00	0.93								
Tl-198m	2	7.79E-02	1	Tl-198	1	2.21E-01	0.47								
Tl-198	2	2.21E-01	0												

Parent Name	Depos Type	Parent HL (days)	No. of Dau	D1 Name	D1 PI	D1 HL (days)	D1 BF	D2 Name	D2 PI	D2 HL (days)	D2 BF	D2 Name	D3 PI	D3 HL (days)	D3 BF
Tl-199	2	3.09E-01	0												
Tl-200	2	1.09E+00	0												
Tl-201	2	3.04E+00	0												
Tl-202	2	1.22E+01	0												
Tl-204	2	1.38E+03	0												
Pb-195m	2	1.10E-02	3	Tl-195	1	4.83E-02	1	Hg-195	2	4.13E-01	1	Au-195	3	1.83E+02	1
Pb-198	2	1.00E-01	1	Tl-198	1	2.21E-01	1								
Pb-199	2	6.25E-02	1	Tl-199	1	3.09E-01	1								
Pb-200	2	8.96E-01	1	Tl-200	1	1.09E+00	1								
Pb-201	2	3.92E-01	1	Tl-201	1	3.40E+00	1								
Pb-202m	2	1.51E-01	1	Tl-202	1	1.22E+01	0.095								
Pb-202	2	1.19E+07	1	Tl-202	1	1.22E+01	0.095								
Pb-203	2	2.17E+00	0												
Pb-205	2	5.22E+09	0												
Pb-209	2	1.36E-01	0												
Pb-210	2	8.14E+03	2	Bi-210	1	5.01E+00	1	Po-210	2	1.38E+02	1				
Pb-211	2	2.51E-02	0												
Pb-211*	2	2.51E-02	0												
Pb-212	2	4.43E-01	1	Bi-212*	1	4.20E-02	1								
Pb-212+	2	4.43E-01	0												
Pb-214	2	1.86E-02	2	Bi-214*	1	1.38E-02	1	Pb-210	2	8.14E+03	0.9998				
Bi-200	2	2.53E-02	2	Pb-200	1	8.96E-01	1	Tl-200	2	1.09E+00	1				
Bi-201	2	7.50E-02	2	Pb-201	1	3.92E-01	1	Tl-201	2	3.04E+00	1				
Bi-202	2	6.96E-02	2	Pb-202m	1	1.51E-01	0.0025	Tl-202	2	1.22E+01	0.095				
Bi-203	2	4.90E-01	1	Pb-203	1	2.17E+00	1								
Bi-205	2	1.53E+01	0												
Bi-206	2	6.24E+00	0												
Bi-207	2	1.39E+04	0												
Bi-210m*	2	1.10E+09	0												
Bi-210	2	5.01E+00	1	Po-210	1	1.38E+02	1								
Bi-212*	2	4.20E-02	0												
Bi-213*	2	3.17E-02	1	Pb-209	1	1.36E-01	0.9784								
Bi-214*	2	1.38E-02	3	Pb-210	1	8.14E+03	1	Bi-210	2	5.01E+00	1	Po-210	3	1.38E+02	1
Po-203	2	2.55E-02	2	Bi-203	1	4.90E-01	0.9989	Pb-203	2	2.17E+00	1				
Po-205	2	7.50E-02	2	Pb-201	1	3.92E-01	0.0014	Tl-201	2	3.04E+00	1	Bi-205	1	1.53E+01	0.9986
Po-207	2	2.43E-01	0												

Parent Name	Depos Type	Parent HL (days)	No. of Dau	D1 Name	D1 PI	D1 HL (days)	D1 BF	D2 Name	D2 PI	D2 HL (days)	D2 BF	D2 Name	D3 PI	D3 HL (days)	D3 BF
Po-210	2	1.38E-02	0												
At-207	2	7.50E-02	3	Bi-203	1	4.90E-01	0.1	Pb-203	2	2.17E+00	1	Po-207	1	2.43E-01	0.9
At-211*	2	3.01E-01	0												
Rn-222*	0	3.82E+00	3	Pb-214	1	1.86E-02	0.9998	Bi-214*	2	1.38E-02	1	Pb-210	3	8.14E+03	1
Rn-222+	0	3.82E+00	3	Pb-210	1	8.14E+03	1	Bi-210	2	5.01E+00	1	Po-210	3	1.38E+02	1
Fr-222*	2	1.00E-02	3	Pb-210	1	8.14E+03	1	Bi-210	2	5.01E+00	1	Po-210	3	1.38E+02	1
Fr-223	2	1.51E-02	2	Ra-223*	1	1.14E+01	1	Pb-211*	2	2.51E-02	1				
Ra-223*	2	1.14E+01	1	Pb-211*	1	2.51E-02	1								
Ra-223+	2	1.14E-01	0												
Ra-224*	2	3.66E+00	2	Pb-212	1	4.43E-01	1	Bi-212*	2	4.20E-02	1				
Ra-224+	2	3.66E+00	0												
Ra-225	2	1.48E+01	2	Ac-225*	1	1.00E+01	1	Bi-213*	2	3.17E-02	1				
Ra-226	2	5.84E+05	0												
Ra-227	2	2.93E-02	1	Ac-227*	1	7.95E+03	1								
Ra-228	2	2.10E+03	3	Ac-228	1	2.55E-01	1	Th-228	2	6.98E+02	1	Ra-224*	3	3.66E+00	1
Ra-228+	2	2.10E+03	3	Th-228	1	6.98E+02	1	Ra-224	2	3.66E+00	1	Pb-212+	3	4.43E-01	1
Ac-224+	2	1.21E-01	3	Ra-224*	1	3.66E+00	0.9	Pb-212	2	4.43E-01	1	Bi-212*	3	4.20E-02	1
Ac-225*	2	1.00E+01	1	Bi-213*	1	3.17E-02	1								
Ac-225+	2	1.00E+01	0												
Ac-226	2	1.21E+00	3	Th-226*	1	2.15E-02	0.828	Pb-210	2	8.14E+03	1	Bi-210	3	5.01E+00	1
Ac-227*	2	7.95E+03	3	Th-227	1	1.87E+01	0.9862	Ra-223*	2	1.14E+01	1	Pb-211*	3	2.51E-02	1
Ac-227+	2	7.95E+03	2	Ra-223*	1	1.14E+01	1	Pb-211*	2	2.51E-02	1				
Ac-228	2	2.55E-01	3	Th-228	1	6.98E+02	1	Pb-212	2	4.43E-01	1	Bi-212*	3	4.20E-02	1
Th-226*	2	2.15E-02	3	Pb-210	1	8.14E+03	1	Bi-210	2	5.01E+00	1	Po-210	3	1.38E+02	1
Th-227	2	1.87E+01	3	Ra-223*	1	1.14E+01	1	Pb-211*	2	2.51E-02	1	Bi-213*	3	3.17E-02	1
Th-227+	2	1.87E+01	0												
Th-228	2	6.98E+02	3	Ra-224*	1	3.66E+00	1	Pb-212	2	4.43E-01	1	Bi-212*	3	4.20E-02	1
Th-228+	2	6.98E+02	0												
Th-229	2	2.68E+06	3	Ra-225	1	1.48E+01	1	Ac-225*	2	1.00E+01	1	Bi-213*	3	3.17E-02	1
Th-230	2	2.81E+07	0												
Th-231	2	1.06E+00	0												
Th-232	2	5.13E+12	3	Ra-228	1	2.10E+03	1	Ac-228	2	2.55E-01	1	Th-228+	3	6.98E+02	1
Th-234*	2	2.41E+01	1	Pa-234	1	2.79E-01	0.0033								
Th-234+	2	2.41E+01	0												
Pa-227*	2	2.66E-02	3	Th-227	1	1.87E+01	0.15	Ra-223*	2	1.14E+01	1	Pb-211*	3	2.51E-02	1
Pa-228	2	9.17E-01	3	Ac-224*	1	1.21E-01	0.02	Ra-224+	2	3.66E+00	0.9	Th-220+	1	6.98E+02	0.98

Parent Name	Depos Type	Parent HL (days)	No. of Dau	D1 Name	D1 PI	D1 HL (days)	D1 BF	D2 Name	D2 PI	D2 HL (days)	D2 BF	D2 Name	D3 PI	D3 HL (days)	D3 BF
Pa-230	2	1.74E+01	3	U-230	1	2.08E+01	0.095	Th-226*	1	2.15E-02	1	Pb-210	3	8.14E+03	1
Pa-231	2	1.20E+01	3	Ac-227*	1	7.95E+03	1	Th-227	2	1.87E+01	1	Ra-223*	3	1.14E+01	1
Pa-232	2	1.31E+00	3	U-232	1	2.63E+04	1	Th-228	2	6.98E+02	1	Ra-224*	3	3.66E+00	1
Pa-233	2	2.70E+07	0												
Pa-234	2	2.79E-01	0												
U-230	2	2.08E+01	2	Th-226*	1	2.15E-02	1	Pb-210	2	8.14E+03	1	Bi-210	3	5.01E+00	1
U-231	2	4.20E+00	0												
U-232	2	2.63E+04	3	Th-228	1	6.98E+02	1	Ra-224*	2	3.66E+00	1	Pb-212+	3	4.43E-01	1
U-233	2	5.79E+07	1	Th-229	1	2.68E+06	1								
U-234	2	8.92E+07	1	Th-230	1	2.81E+07	1								
U-235	2	2.57E+11	1	Th-231	1	1.06E+00	1								
U-236	2	8.55E+09	0												
U-237	2	6.75E+00	0												
U-238	2	1.63E+12	1	Th-234*	1	2.41E+01	1								
U-239	2	1.63E-02	1	Np-239	1	2.36E+00	1								
U-240*	2	5.88E-01	1	Pu-240	1	2.39E+06	1								
Np-232	2	1.02E-02	3	U-232	1	2.63E+04	1	Th-228	2	6.98E+02	1	Ra-224+	3	3.66E+00	1
Np-233	2	2.51E-02	0												
Np-234	2	4.40E+00	0												
Np-235	2	3.96E+02	0												
Np-236	2	4.20E+07	3	Pu-236	1	1.04E+03	0.089	U-232	2	2.63E+04	1	Th-228+	3	6.98E+02	1
Np-236m	2	9.38E-01	3	Pu-236	1	1.04E+03	0.48	U-232	2	2.63E+04	1	Th-228+	3	6.98E+02	1
Np-237	2	7.81E+08	1	Pa-233	1	2.70E+01	1								
Np-238	2	2.12E+00	1	Pu-238	1	3.20E+04	1								
Np-239	2	2.36E+00	1	Pu-239	1	8.78E+06	1								
Np-240	2	4.51E-02	0												
Pu-234	2	3.67E-01	3	U-230	1	2.08E+01	0.06	Th-226*	2	2.15E-02	1	Np-234	1	4.40E+00	0.94
Pu-235	2	1.76E-02	1	Np-235	1	3.96E+02	1								
Pu-236	2	1.04E+03	3	U-232	1	2.63E+04	1	Th-228	2	6.98E+02	1	Ra-224+	3	3.66E+00	1
Pu-237	2	4.53E-01	0												
Pu-238	2	3.20E+04	0												
Pu-239	2	8.78E+06	0												
Pu-240	2	2.39E+06	0												
Pu-241	2	5.26E+03	1	Am-241	1	1.58E+05	1								
Pu-242	2	1.37E+08	0												
Pu-243	2	2.07E-01	0												

Parent Name	Depos Type	Parent HL (days)	No. of Dau	D1 Name	D1 PI	D1 HL (days)	D1 BF	D2 Name	D2 PI	D2 HL (days)	D2 BF	D2 Name	D3 PI	D3 HL (days)	D3 BF
Pu-244	2	3.01E+10	1	U-240*	1	5.88E-01	1								
Pu-244+	2	3.01E+10	0												
Pu-245	2	4.38E-01	1	Am-245	1	8.54E-02	1								
Pu-246	2	1.09E+01	1	Am-246	1	2.71E-02	1								
Am-237	2	5.07E-02	0												
Am-238	2	6.81E-02	1	Pu-238	1	3.20E+04	1								
Am-239	2	4.96E-01	0												
Am-240	2	2.12E+00	0												
Am-241	2	1.58E+05	0												
Am-242m	2	5.55E+04	3	Am-242	1	6.68E-01	1	Cm-242	2	1.63E+02	0.827	Np-238	1	2.12E+00	0.0048
Am-242	2	6.68E-01	1	Cm-242	1	1.63E+02	0.827								
Am-243	2	2.69E+06	1	Np-239	1	2.36E+00	1								
Am-244m	2	1.81E-02	1	Cm-244	1	6.61E+03	1								
Am-244	2	4.21E-01	1	Cm-244	1	6.61E+03	1								
Am-245	2	8.54E-02	0												
Am-246m	2	1.74E-02	0												
Am-246	2	2.71E-02	0												
Cm-238	2	1.00E-01	3	Pu-234	1	3.67E-01	0.1	Np-234	2	4.40E+00	0.94	Am-238	1	6.81E-02	0.9
Cm-240	2	2.70E+01	2	Pu-236	1	1.04E+03	1	U-232	2	2.63E+04	1				
Cm-241	2	3.28E+01	2	Pu-237	1	4.53E+01	0.01	Am-241	1	1.58E+05	0.99				
Cm-242	2	1.63E+02	1	Pu-238	1	3.20E+04	1								
Cm-243	2	1.04E+04	0												
Cm-244	2	6.61E+03	0												
Cm-245	2	3.10E+06	0												
Cm-246	2	1.73E+06	0												
Cm-247	2	5.69E+09	1	Pu-243	1	2.07E-01	1								
Cm-247+	2	5.69E+09	0												
Cm-248	2	1.24E+08	0												
Cm-249	2	4.45E-02	2	Bk-249	1	3.20E+02	1	Cf-249	2	1.28E+05	1				
Cm-250	2	3.03E+06	3	Pu-246	1	1.09E+01	0.25	Am-246	2	2.71E-02	1	Bk-250	1	1.34E-01	0.14
Bk-245	2	4.94E+00	0												
Bk-246	2	1.83E+00	0												
Bk-247	2	5.04E+05	0												
Bk-249	2	3.20E+02	1	Cf-249	1	1.28E+05	1								
Bk-250	2	1.34E-01	1	Cf-250	1	4.77E+03	1								
Cf-244	2	1.35E-02	3	Cm-240	1	2.70E+01	1	Pu-236	2	1.04E+03	1	U-232	3	2.63E+04	1

Parent Name	Depos Type	Parent HL (days)	No. of Dau	D1 Name	D1 PI	D1 HL (days)	D1 BF	D2 Name	D2 PI	D2 HL (days)	D2 BF	D2 Name	D3 PI	D3 HL (days)	D3 BF
Cf-246	2	1.49E+00	2	Cm-242	1	1.63E+02	0.9997	Pu-238	2	3.20E+04	1				
Cf-248	2	3.34E+02	1	Cm-244	1	6.61E+03	1								
Cf-249	2	1.28E+05	0												
Cf-250	2	4.77E+03	0												
Cf-251	2	3.28E+05	0												
Cf-252	2	9.63E+02	0												
Cf-253	2	1.78E+01	3	Es-253	1	2.05E+01	1	Bk-249	2	3.20E+02	1	Cf-249	3	1.28E+05	1
Cf-254	2	6.05E+01	0												
Es-250	2	8.75E-02	1	Cf-250	1	4.77E+03	1								
Es-251	2	1.38E+00	0												
Es-253	2	2.05E+01	2	Bk-249	1	3.20E+02	1	Cf-249	2	1.28E+05	1				
Es-254m	2	1.64E+00	3	Fm-254	1	1.35E-01	0.98	Cf-250	2	4.77E+03	1	Bk-250	1	1.34E-01	0.0032
Es-254	2	2.76E+02	1	Bk-250	1	1.34E-01	1								
Es-254+	2	2.76E+02	0												
Fm-252	2	9.46E-01	2	Cf-248	1	3.34E+02	1	Cm-244	1	6.61E+03	1				
Fm-253	2	3.00E+00	3	Es-253	1	2.05E+01	0.88	Bk-249	2	3.20E+02	1	Cf-249	3	1.28E+05	1
Fm-254	2	1.35E-01	1	Cf-250	1	4.77E+03	1								
Fm-255	2	8.36E-01	0												
Fm-257	2	1.01E+02	3	Cf-253	1	1.78E+01	0.9979	Es-253	2	2.05E+01	1	Bk-249	3	3.20E+02	1
Md-257	2	2.17E-01	3	Fm-257	1	1.01E+02	0.9	Cf-253	2	1.78E+01	0.9979	Es-253	1	2.05E+01	0.1
Md-258	2	5.50E+01	2	Es-254	1	2.76E+02	1	Bk-250	2	1.34E-01	1				
U-Natrl	2	8.92E+07	0												
U-Enrch	2	1.63E+12	0												
UF$_6$	2	1.63E+12	0												

NRC FORM 335
(12-2010)
NRCMD 3.7

U.S. NUCLEAR REGULATORY COMMISSION

BIBLIOGRAPHIC DATA SHEET

(See instructions on the reverse.)

1. REPORT NUMBER (Assigned by NRC, Add Vol., Supp., Rev., and Addendum Numbers, if any.)	NUREG 1940

2. TITLE AND SUBTITLE

RASCAL 4: Description of Models and Methods

3. DATE REPORT PUBLISHED	
MONTH	YEAR
December	2012

4. FIN OR GRANT NUMBER
R1174

5. AUTHOR(S)

J. V. Ramsdell, Jr., Pacific Northwest Laboratories
George F. Athey, Athey Consulting
Stephen McGuire, (retired)
Lou Brandon

6. TYPE OF REPORT

Final Technical Report

7. PERIOD COVERED (Inclusive Dates)

NA

8. PERFORMING ORGANIZATION - NAME AND ADDRESS (If NRC, provide Division, Office or Region, U. S. Nuclear Regulatory Commission, and mailing address; if contractor, provide name and mailing address.)

Division of Preparedness and Response
Office of Nuclear Security and Incident Response
U.S. Nuclear Regulatory Commission
Washington, DC 20555-0001

9. SPONSORING ORGANIZATION - NAME AND ADDRESS (If NRC, type "Same as above", if contractor, provide NRC Division, Office or Region, U. S. Nuclear Regulatory Commission, and mailing address.)

Same as above

10. SUPPLEMENTARY NOTES
Supercedes NUREG 1887

11. ABSTRACT (200 words or less)

RASCAL 4 is a significant advancement in the U.S. Nuclear Regulatory Commission's emergency response consequence assessment tools. RASCAL 4 includes improvements in the models and methods related to source term calculations, atmospheric dispersion and deposition, and dose calculations. Changes to the user interface will facilitate data entry, processing, and analysis. This report describes the models and methods that are included in RASCAL 4, and it describes the consequence assessment implications of changes in models and methods from RASCAL 3.

12. KEY WORDS/DESCRIPTORS (List words or phrases that will assist researchers in locating the report.)

RASCAL, emergency response, dose assessment, dose projections, atmospheric dispersion, plume modeling

13. AVAILABILITY STATEMENT
unlimited

14. SECURITY CLASSIFICATION
(This Page)
unclassified
(This Report)
unclassified

15. NUMBER OF PAGES

16. PRICE

NRC FORM 335 (12-2010)

Printed
on recycled
paper

Federal Recycling Program

NUREG-1940

RASCAL 4: Description of Models and Methods

December 2012

www.ingramcontent.com/pod-product-compliance
Lightning Source LLC
Chambersburg PA
CBHW080241180526
45167CB00006B/2365